Henry Augustus Rowland

On the Mechanical Equivalent of Heat

With subsidiary researches on the variation of the mercurial from the air thermometer, and on the variation of the specific heat of water

Henry Augustus Rowland

On the Mechanical Equivalent of Heat
With subsidiary researches on the variation of the mercurial from the air thermometer, and on the variation of the specific heat of water

ISBN/EAN: 9783337012854

Printed in Europe, USA, Canada, Australia, Japan

Cover: Foto ©berggeist007 / pixelio.de

More available books at **www.hansebooks.com**

ON THE

MECHANICAL EQUIVALENT OF HEAT,

WITH SUBSIDIARY RESEARCHES ON

THE VARIATION OF THE MERCURIAL FROM THE AIR
THERMOMETER, AND ON THE VARIATION OF
THE SPECIFIC HEAT OF WATER.

BY

HENRY A. ROWLAND,

PROFESSOR OF PHYSICS IN THE JOHNS HOPKINS UNIVERSITY.

Presented June 11th, 1879.

[Reprinted from the Proceedings of the American Academy of
Arts and Sciences.]

CAMBRIDGE:
UNIVERSITY PRESS: JOHN WILSON & SON.
1880.

INVESTIGATIONS ON LIGHT AND HEAT, made and published wholly or in part with appropriation from the RUMFORD FUND.

V.

ON THE MECHANICAL EQUIVALENT OF HEAT, WITH SUBSIDIARY RESEARCHES ON THE VARIATION OF THE MERCURIAL FROM THE AIR THERMOMETER, AND ON THE VARIATION OF THE SPECIFIC HEAT OF WATER.

BY HENRY A. ROWLAND,[*]
Professor of Physics in the Johns Hopkins University.

Presented June 11th, 1879.

CONTENTS.

I. Introductory Remarks	75
II. Thermometry	77
(a.) General View of Thermometry	77
(b.) The Mercurial Thermometer	78
(c.) Relation of the Mercurial and Air Thermometers	83
1. General and Historical Remarks	83
2. Description of Apparatus	90
3. Results of Comparison	97
(d.) Reduction to the Absolute Scale	112
Appendix to Thermometry	116
III. Calorimetry	119
(a.) Specific Heat of Water	119
(b.) Heat Capacity of the Calorimeter	131
IV. Determination of Equivalent	137
(a.) Historical Remarks	137
1. General Review of Methods	137
2. Results of Best Determinations	140
(b.) Description of Apparatus	155
1. Preliminary Remarks	155
2. General Description	157
3. Details	158
(c.) Theory of the Experiment	168
1. Estimation of Work done	162
2. Radiation	168
3. Corrections to Thermometers, etc.	171
(d.) Results	173
1. Constant Data	173
2. Experimental Data and Tables of Results	174
V. Concluding Remarks, and Criticism of Results and Methods	197

I. — INTRODUCTORY REMARKS.

AMONG the more important constants of nature, the ratio of the heat unit to the unit of mechanical work stands forth prominent, and is used almost daily by the physicist. Yet, when we come to consider

---

[*] This research was originally to have been performed in connection with Professor Pickering, but the plan was frustrated by the great distance between our residences. An appropriation for this experiment was made by the American Academy of Arts and Sciences at Boston, from the fund which was instituted by Count Rumford, and liberal aid was also given by the Trustees of the Johns Hopkins University, who are desirous, as far as they can, to promote original scientific investigation.

the history of the subject carefully, we find that the only experimenter who has made the determination with anything like the accuracy demanded by modern science, and by a method capable of giving good results, is Joule, whose determination of thirty years ago, confirmed by some recent results, to-day stands almost, if not quite, alone among accurate results on the subject.

But Joule experimented on water of one temperature only, and did not reduce his results to the air thermometer; so that we are still left in doubt, even to the extent of one per cent, as to the value of the equivalent on the air thermometer.

The reduction of the mercurial to the air thermometer, and thence to the absolute scale, has generally been neglected between 0° and 100° by most physicists, though it is known that they differ several tenths of a degree at the 45° point. In calorimetric researches this may produce an error of over one, and even approaching two per cent, especially when a Geissler thermometer is used, which is the worst in this respect of any that I have experimented on; and small intervals on the mercurial thermometers differ among themselves more than one per cent from the difference of the glass used in them.

Again, as water is necessarily the liquid used in calorimeters, its variation of specific heat with the temperature is a very important factor in the determination of the equivalent. Strange as it may appear, we may be said to know almost nothing about the variation of the specific heat of water with the temperature between 0° and 100° C.

Regnault experimented only above 100° C. The experiments of Hirn, and of Jamin and Amaury, are absurd, from the amount of variation which they give. Pfaundler and Plattner confined themselves to points between 0° and 13°. Münchausen seems to have made the best experiments, but they must be rejected because he did not reduce to the air thermometer.

In the present series of researches, I have sought, firstly, a method of measuring temperatures on the perfect gas thermometer with an accuracy scarcely hitherto attempted, and to this end have made an extended study of the deviation of ordinary thermometers from the air thermometer; and, secondly, I have sought a method of determining the mechanical equivalent of heat so accurate, and of so extended a range, that the variation of the specific heat of water should follow from the experiments alone.

As to whether or not these have been accomplished, the following pages will show. The curious result that the specific heat of water on the *air* thermometer decreases from 0° to about 30° or 35°, after which it *increases*, seems to be an entirely unique fact in nature, seeing that there is apparently no other substance hitherto experimented upon whose specific heat *decreases* on rise of temperature without change of state. From a thermodynamic point of view, however, it is of the same nature as the decrease of specific heat which takes place after the vaporization of a liquid.

The close agreement of my result at 15°.7 C. with the old result of Joule, after approximately reducing his to the air thermometer and latitude of Baltimore, and correcting the specific heat of copper, is very satisfactory to us both, as the difference is not greater than 1 in 400, and is probably less.

I hope at some future time to make a comparison with Joule's thermometers, when the difference can be accurately stated.

II.—THERMOMETRY.

(a.) General View.

The science of thermometry, as ordinarily studied, is based upon the changes produced in bodies by heat. Among these we may mention change in volume, pressure, state of aggregation, dissociation, amount and color of light reflected, transmitted, or emitted, hardness, pyro-electric and thermo-electric properties, electric conductivity or specific induction capacity, magnetic properties, thermo-dynamic properties, &c.; and on each of these may be based a system of thermometry, each one of which is perfect in itself, but which differs from all the others widely. Indeed, each method may be applied to nearly all the bodies in nature, and hundreds or thousands of thermometric scales may be produced, which may be made to agree at two fixed points, such as the freezing and boiling points of water, but which will in general differ at nearly, if not all, other points.

But from the way in which the science has advanced, it has come to pass that all methods of thermometry in general use to the present time have been reduced to two or three, based respectively on the apparent expansion of mercury in glass and on the absolute expansion of some gas, and more lately on the second law of thermodynamics.

Each of these systems is perfectly correct in itself, and we have no right to designate either of them as incorrect. We must decide *a priori* on some system, and then express all our results in that system: the accuracy of science demands that there should be no ambiguity on that subject. In deciding among the three systems, we should be guided by the following rules: —

1st. The system should be perfectly definite, so that the same temperature should be indicated, whatever the thermometer.

2d. The system should lead to the most simple laws in nature.

Sir William Thomson's absolute system of thermometry, coinciding with that based on the expansion of a perfect gas, satisfies these most nearly. The mercurial thermometer is not definite unless the kind of glass is given, and even then it may vary according to the way the bulb is blown. The gas thermometer, unless the kind of gas is given, is not definite. And, further, if the temperature as given by either of these thermometers was introduced into the equations of thermodynamics, the simplest of them would immediately become complicated.

Throughout a small range of temperature, these systems agree more or less completely, and it is the habit even with many eminent physicists to regard them as coincident between the freezing and boiling points of water. We shall see, however, that the difference between them is of the highest importance in thermometry, especially where differences of temperature are to be used.

For these reasons I have reduced all my measures to the absolute system.

The relation between the absolute system and the system based on the expansion of gases has been determined by Joule and Thomson in their experiments on the flow of gases through porous plugs (Philosophical Transactions for 1862, p. 579). Air was one of the most important substances they experimented upon.

To measure temperature on the absolute scale, we have thus only to determine the temperature on the air thermometer, and then reduce to the absolute scale. But as the air thermometer is very inconvenient to use, it is generally more convenient to use a mercurial thermometer which has been compared with the air thermometer. Also, for small changes of temperature the air thermometer is not sufficiently sensitive, and a mercurial thermometer is necessary for interpolation. I shall occupy myself first with a careful study of the mercurial thermometer.

(b.) **The Mercurial Thermometer.**

Of the two kinds of mercurial thermometers, the weight thermometer is of little importance to our subject. I shall therefore confine myself principally to that form having a graduated stem. For convenience in use and in calibration, the principal bulb should be elongated, and another small bulb should be blown at the top. This latter is also of the utmost importance to the accuracy of the instrument, and is placed there by nearly all makers of standards.* It is used to place some of the mercury in while calibrating, as well as when a high temperature is to be measured; also, the mercury in the larger bulb can be made free from air-bubbles by its means.

Most standard thermometers are graduated to degrees; but Regnault preferred to have his thermometers graduated to parts of equal capacity whose value was arbitrary, and others have used a single millimeter division. As thermometers change with age, the last two methods are the best; and of the two I prefer the latter where the highest accuracy is desired, seeing that it leaves less to the maker and more to the scientist. The cross-section of the tube changes continuously from point to point, and therefore the distribution of marks on the tube should be continuous, which would involve a change of the dividing engine for each division. But as the maker divides his tube, he only changes the length of his divisions every now and then, so as to average his errors. This gives a sufficiently exact graduation for large ranges of temperature; but for small, great errors may be introduced. Where there is an arbitrary scale of millimeters, I believe it is possible to calibrate the tube so that the errors shall be less than can be seen with the naked eye, and that the table found shall represent very exactly the gradual variation of the tube.

In the calibration of my thermometers with the millimetric scale, I have used several methods, all of which are based upon some graphical method. The first, which gives all the irregularities of the tube with great exactness, is as follows.

A portion of the mercury having been put in the upper bulb, so as to leave the tube free, a column about 15^{mm} long is separated off. This is moved from point to point of the tube, and its length carefully measured on the dividing engine. It is not generally necessary to move the column its own length every time, but it may be moved 20^{mm} or 25^{mm}, a record of the position of its centre being kept. To eliminate any errors of division or of the dividing engine, readings were then taken on the scale, and the lengths reduced to their value in scale divisions. The area of the tube at every point is inversely as the length of the column. We shall thus have a series of figures nearly equal to each other, if the tube is good. By subtracting the

---

* Geissler and Casella omit it, which should condemn their thermometers.

smallest from each of the others, and plotting the results as ordinates, with the thermometer scale as abscissas, and drawing a curve through the points so found, we have means of finding the area at any point. The curve should not be drawn exactly through the points, but rather around them, seeing they are the average areas for some distance each side of the point. With good judgment, the curve can be drawn with great accuracy. I then draw ordinates every 10^{mm}, and estimate the average area of the tube for that distance, which I set down in a table. As the lengths are uniform, the volume of the tube to any point is found by adding up the areas to that point.

But it would be unwise to trust such a method for very long tubes, seeing the mercury column is so short, and the columns are not end to end. Hence I use it only as supplementary to one where the column is about 50^{mm} long, and is always moved its own length. This establishes the volumes to a series of points about 50^{mm} apart, and the other table is only used to interpolate in this one. There seems to be no practical object in using columns longer than this.

Having finally constructed the arbitrary table of volumes, I then test it by reading with the eye the length of a long mercury column. No certain error was thus found at any point of any of the thermometers which I have used in these experiments.

While measuring the column, great care must be taken to preserve all parts of the tube at a uniform temperature, and only the extreme ends must be touched with the hands, which should be covered with cloth.

If V is the volume on this arbitrary scale, the temperature on the mercurial thermometer is found from the formula $T = CV - t_0$, where C and t_0 are constants to be determined. If the thermometer contains the 0° and 100° points, we have simply

$$C = \frac{100}{V_{100} - V_0}.$$

Otherwise C is found by comparison with some other thermometer, which must be of the same kind of glass.

It is to be carefully noted that the temperature on the mercurial thermometer, as I have defined it, is proportional to the apparent expansion of mercury as measured on the stem. By defining it as proportional to the true volume of mercury in the stem, we have to introduce a correction to ordinary thermometers, as Poggendorf has shown. As I only use the mercurial thermometer to compare with the air thermometer, and as either definition is equally correct, I will not further discuss the matter, but will use the first definition, as being the simplest.

In the above formula I have implicitly assumed that the apparent expansion is only a function of the temperature; but in solid bodies like glass there seems to be a progressive change in the volume as time advances, and especially after it has been heated. And hence in mercurial and alcohol thermometers, and probably in general in all thermometers which depend more or less on the expansion of solid bodies, we find that the reading of the thermometer depends, not only on its present temperature, but also on that to which it has been subjected within a short time; so that, on heating a thermometer up to a certain temperature, it does not stand at the same point as if it had been cooled from a higher temperature to the given temperature. As these effects are without doubt due to the glass envelope, we might greatly diminish them by using thermometers filled with liquids which expand more than mercury: there are many of these which expand six or eight times as much, and so the irregularity might be diminished in this ratio. But in this case we should find that the correction for that part of the stem which was outside the vessel whose temperature we were determining would be increased in the same proportion; and besides, as all the liquids are quite volatile, or at least wet the glass, there would be an irregularity introduced on that account. A thermometer with liquid in the bulb and mercury in the stem would obviate these inconveniences; but even in this case the stem would have to be calibrated before the thermometer was made. By a comparison with the air-thermometer, a proper formula could be obtained for finding the temperature.

But I hardly believe that any thermometer superior to the mercurial can at present be made, — that is, any thermometer within the same compass as a mercurial thermometer, — and I think that the best result for small ranges of temperature can be obtained with it by studying and avoiding all its sources of error.

To judge somewhat of the laws of the change of zero within the limits of temperature which I wished to use, I took thermometer No. 6163, which had lain in its case during four months at an average temperature of about 20° or 25° C., and observed the zero point, after heating to various temperatures, with the following result. The time of heating was only a few minutes, and the zero point was taken immediately after; some fifteen minutes, however, being necessary for the thermometer to entirely cool.

TABLE I. — Showing Change of Zero Point.

Temperature of Bulb before finding the 0 Point.	Change of 0 Point.	Temperature of Bulb before finding the 0 Point.	Change of 0 Point.
22°.5	0	70°.0	—.115
30.0	—.016	81.0	—.170
40.5	—.083	90.0	—.231
51.0	—.039	100.0	—.313
60.0	—.105	100.0	—.347

The second 100° reading was taken after boiling for some time.

It is seen that the zero point is always lower after heating, and that in the limits of the table the lowering of the zero is about proportional to the square of the increase of temperature above 25° C. This law is not true much above 100°, and above a certain temperature the phenomenon is reversed, and the zero point is higher after heating; but for the given range it seems quite exact.

It is not my purpose to make a complete study of this phenomenon with a view to correcting the thermometer, although this has been undertaken by others. But we see from the table that the error cannot exceed certain limits. The range of temperature which I have used in each experiment is from 20° to 30° C., and the temperature rarely rose above 40° C. The change of zero in this range only amounts to 0°.03 C.

The exact distribution of the error from this cause throughout the scale has never been determined, and it affects my results so little that I have not considered it worth investigating. It seems probable, however, that the error is distributed throughout the scale. If it were uniformly distributed, the value of each division would be less than before by the ratio of the lowering at zero to the temperature to which the thermometer was heated.

The maximum errors produced in my thermometers by this cause would thus amount to 1 in 1300 nearly for the 40° thermometer, and to about 1 in 2000 for the others. Rather than allow for this, it is better to allow time for the thermometer to resume its original state.

Only a few observations were made upon the rapidity with which the zero returned to its original position. After heating to 81°, the zero returned from —0°.170 to —0°.148 in two hours and a half. After heating to 100°, the zero returned from —0°.347 to —0°.110 in nine days, and to —0°.022 in one month. Reasoning from this, I should say that in one week thermometers which had not been heated above 40° should be ready for use again, the error being then supposed to be less than 1 in 4000, and this would be partially eliminated by comparing with the air thermometer at the same intervals as the thermometer is used, or at least heating to 40° one week before comparing with the air thermometer.

As stated before, when a thermometer is heated to a very high point, its zero point is raised instead of lowered, and it seems probable that at some higher point the direction of change is reversed again; for, after the instrument comes from the maker, the zero point constantly rises until it may be 0°.6 above the mark on the tube. This gradual change is of no importance in my experiments, as I only use differences of temperature, and also as it was almost inappreciable in my thermometers.

Another source of error in thermometers is that due to the pressure on the bulb. In determining the freezing point, large errors may be made, amounting to several hundredths of a degree, by the pressure of pieces of ice. In my experiments, the zero point was determined in ice, and then the thermometer was immersed in the water of the comparator at a depth of about 60^{cm}. The pressure of this water affected the thermometer to the extent of about 0°.01, and a correction was accordingly made. As differences of temperature were only needed, no correction was made for variation in pressure of the air.

It does not seem to me well to use thermometers with too small a stem, as I have no doubt that they are subject to much greater irregularities than those with a coarse bore. For the capillary action always exerts a pressure on the bulb. Hence, when the mercury rises, the pressure is due to a *rising* meniscus which causes greater pressure than the *falling* meniscus. Hence, an *apparent friction* of the mercurial column. Also, the capillary constant of mercury seems to depend on the electric potential of its surface, which may not be constant, and would thus cause an irregularity.

My own thermometers did not show any apparent action of this kind, but Pfaundler and Plattner mention such an action, though they give another reason for it.

(c.) Relation of the Mercurial and Air Thermometers.

1. General and Historical Remarks.

Since the time of Dulong and Petit, many experiments have been made on the difference between the mercurial and the air thermometer, but unfortunately most of them have been at high temperatures. As weight thermometers have been used by some of the best experimenters, I shall commence by proving that the weight thermometer and stem thermometer give the same temperature; at the same time, however, obtaining a convenient formula for the comparison of the air thermometer with the mercurial.

For the expansion of mercury and of glass the following formulæ must hold: —

For mercury, $V = V_0 (1 + at + bt^2 + \&c.)$;
" glass, $V' = V'_0 (1 + \alpha t + \beta t^2 + \&c.)$.

In both the weight and stem thermometers we must have $V = V'$.

$$\therefore V'_0 = V_0 \frac{1 + at + bt^2 + \&c.}{1 + \alpha t + \beta t^2 + \&c.} = V_0 (1 + At + Bt^2 + \&c.),$$

where V'_0 and V_0 are the volumes of the glass and of the mercury reduced to zero, and t is the temperature on the air thermometer. The temperature by the weight thermometer is

$$T = 100 \frac{P_0 - P_t}{P_0 - P_{100}} \frac{P_{100}}{P_t} = 100 \frac{\frac{P_0}{P_t} - 1}{\frac{P_0}{P_{100}} - 1},$$

where P_0, P_t, &c. are the weights of mercury in the bulb at 0° C., t° C., &c.

Now these weights are directly as the volumes of the mercury at 0°.

$$\therefore \frac{P_0}{P_t} = 1 + At + Bt^2 + \&c.,$$

seeing that V is constant.

$$\therefore T = 100 \frac{At + Bt^2 + \&c.}{100 A + (100)^2 B + \&c.}.$$

In the stem thermometers we have V_0, the volume of mercury at 0°, constant, and the volume of the glass that the mercury fills, reduced to 0°, variable. As the volume of the glass V'_0 is the volume reduced to 0°, it will be proportional to the volume of bulb plus the volume of the tube *as read off on the scale* which should be on the tube.

$$T = 100 \frac{(V'_0)_t - (V'_0)_0}{(V'_0)_{100} - (V'_0)_0} = 100 \frac{\frac{(V'_0)_t}{(V'_0)_0} - 1}{\frac{(V'_0)_{100}}{(V'_0)_0} - 1};$$

$$\therefore \quad T = 100 \, \frac{At + Bt^2 + \&c.}{100\,A + (100)^2\,B^2 + \&c.},$$

which is the same as for the weight thermometer.

If the fixed points are 0° and t' instead of 0° and 100°, we can write

$$T = t' \, \frac{At + Bt^2 + Ct^3 + \&c.}{At' + Bt'^2 + Ct'^3 + \&c.}$$

$$T = t \left\{ 1 + (t - t') \left\{ \frac{B}{A} + \frac{C}{A} + \frac{B^2}{A^2} t' + \frac{C}{A} t \right\} + \&c. \right\}$$

$$T = t \left\{ 1 + (t - t') \left[\frac{B}{A} + \frac{B^2}{A^2} t' + \frac{C}{A} (t + t') \right] + \&c. \right\}$$

As T and t are nearly equal, and as we shall determine the constants experimentally, we may write

$$t = T - at(t' - t)(b - t) + \&c.,$$

where t is the temperature on the air thermometer, and T that on the mercurial thermometer, and a and b are constants to be determined for each thermometer.

The formula might be expanded still further, but I think there are few cases which it will not represent as it is. Considering b as equal to 0, a formula is obtained which has been used by others, and from which some very wrong conclusions have been drawn. In some kinds of glass there are three points which coincide with the air thermometer, and it requires at least an equation of the third degree to represent this.

The three points in which the two thermometers coincide are given by the roots of the equation

$$t(t' - t)(b - t) = 0,$$

and are, therefore,

$$t = 0$$
$$t = t'$$
$$t = b.$$

In the following discussion of the historical results, I shall take 0° and 100° as the fixed points. Hence, $t' = 100°$. To obtain a and b, two observations are needed at some points at a distance from 0° and 100°. That we may get some idea of the values of the constants in the formula for different kinds of glass, I will discuss some of the experimental results of Regnault and others with this in view.

Regnault's results are embodied, for the most part, in tables given on p. 239 of the first volume of his *Relation des Expériences*. The figures given there are obtained from curves drawn to represent the mean of his experiments, and do not contain any theoretical results. The direct application of my formula to his experiments could hardly be made without immense labor in finding the most probable value of the constants.

But the following seem to satisfy the experiments quite well : —

Cristal de Choisy-le-Roi	$b = 0$,	$a = .000\,000\,32$.
Verre Ordinaire	$b = 245°$,	$a = .000\,000\,34$.
Verre Vert	$b = 270°$,	$a = .000\,000\,095$.
Verre de Suède	$b = +10°$,	$a = .000\,000\,14$.

From these values I have calculated the following : —

TABLE II. — REGNAULT'S RESULTS COMPARED WITH THE FORMULA.

Air Thermom.	Choisy-le-Roi.			Verre Ordinaire.			Verre Vert.			Verre de Suède.		
	Observed.	Calculated.	Difference.	Observed.	Calculated.	Difference.	Observed.	Calculated.	Difference.	Observed.	Calculated.	Difference.
100	0	0	0	0	0	0	0	0	0	0	0	0
120	120.12	120.09	+.03	119.95	119.90	+.05	120.07	120.09	—.01	120.04	120.04	0
140	140.29	140.25	+.04	139.85	139.80	+.05	140.21	140.22	—.01	140.11	140.10	+.01
160	160.52	160.49	+.03	159.74	159.72	+.02	160.40	160.39	+.01	160.20	160.21	—.01
180	180.80	180.83	—.03	179.63	179.68	—.05	180.60	180.62	—.02	180.33	180.34	—.01
200	201.25	201.28	—.03	199.70	199.69	+.01	200.80	200.89	—.09	200.50	200.53	—.03
220	221.82	221.86	—.04	219.80	219.78	+.02	221.20	221.23	—.03	220.75	220.78	—.03
240	242.55	242.56	—.01	239.90	239.96	—.06	241.60	241.63	—.03	241.16	241.08	+.08
260	263.44	263.46	—.02	260.20	260.21	—.01	262.15	262.09	+.07			
280	284.48	284.52	—.04	*280.58	280.60	— .02	282.85	282.63	+.22			
300	305.72	305.76	—.04	301.08	301.12	—.04						
320	327.25	327.20	—.05	321.80	321.80	00						
340	349.30	348.88	+.42	434.00	342.64	+.36						

The formula, as we see from the table, represents all Regnault's curves with great accuracy, and if we turn to his experimental results we shall find that the deviation is far within the limits of the experimental errors. The greatest deviation happens at 340°, and may be accounted for by an error in drawing the curve, as there are few experimental results so high as this, and the formula seems to agree with them almost as well as Regnault's own curve.

The object of comparing the formula with Regnault's results at temperatures so much higher than I need, is simply to test the formula through as great a range of temperatures, and for as many kinds of

---

* Corrected from 280.52 in Regnault's table.

glass, as possible. If it agrees reasonably well throughout a great range, it will probably be very accurate for a small range, provided we obtain the constants to represent that small range the best.

Having obtained a formula to represent any series of experiments, we can hardly expect it to hold for points outside our series, or even for interpolating between experiments too far apart, as, very often, a small change in one of the constants may affect the part we have not experimented on in a very marked manner. Thus in applying the formula to points between 0° and 100° the value of b will affect the result very much. In the case of the glass Choisy-le-Roi many values of b will satisfy the observations besides $b = 0$. For the ordinary glass, however, b is well determined, and the formula is of more value between 0° and 100°.

The following table gives the results of the calculation.

TABLE III. — REGNAULT'S RESULTS COMPARED WITH THE FORMULA.

Air Thermometer.	Calculated $a = .000\ 000\ 32$ $b = 0.$	Calculated $a = .000\ 000\ 34$ $b = 245.$	Observed.	Δ	Calculated $a = .000\ 000\ 44$ $b = 260.$	Δ
	Choisy-le-Roi.	Verre Ordinaire.	Verre Ordinaire.		Verre Ordinaire.	
0	0	0	0	....	0	....
10	10.00	10.07	....	....	10.10	....
20	19.99	20.12	....	....	20.17	....
30	29.98	30.15	30.12	+.03	30.21	+.09
40	39.97	40.17	40.23	−.06	40.23	0
50	49.96	50.17	50.23	−.06	50.23	0
60	59.95	60.15	60.24	−.09	60.21	−.03
70	69.95	70.12	70.22	−.10	70.18	−.04
80	79.96	80.09	80.10	−.01	80.11	+.01
90	89.97	90.05	....	....	90.07	....
100	100	100	100	....	100	....

Regnault does not seem to have published any experiments on Choisy-le-Roi glass between 0° and 100°, but in the table between pp. 226, 227, there are some results for ordinary glass. The separate observations do not seem to have been very good, but by combining the total number of observations I have found the results given above. The numbers in the fourth column are found by taking the mean of Regnault's results for points as near the given temperature as possible. The agreement is only fair, but we must remember that the same specimens of glass were not used in this experiment as in the others, and that for these specimens the agreement is also poor above 100°. The values $a = .000\ 000\ 44$ and $b = 260°$ are much better for these specimens, and the seventh column contains the values calculated from these values. These values also satisfy the observations above 100° for the given specimens.

The table seems to show that between 0° and 100° a thermometer of Choisy-le-Roi almost exactly agrees with the air thermometer. But this is not at all conclusive. Regnault, however, remarks,* that between 0° and 100° thermometers of this glass agree more nearly with the air thermometer than those of ordinary glass, though he states the difference to amount to .1 to .2 of a degree, the mercurial thermometer standing *below* the air thermometer. With the exception of this remark of Regnault's, no experiments have ever been published in which the direction of the deviation was similar to this. All experimenters have found the mercurial thermometer to stand *above* the air thermometer between 0° and 100°, and my own experiments agree with this. However, no general rule for all kinds of glass can be laid down.

Boscha has given an excellent study of Regnault's results on this subject, though I cannot agree with all his conclusions on this subject. In discussing the difference between 0 and 100° he uses a formula of the form

$$T - t = \frac{b}{a} t (100 - t),$$

and deduces from it the erroneous conclusion that the difference is greatest at 50° C., instead of between 40° and 50°. His results for $T - t$ at 50° are

Choisy-le-Roi	—.22
Verre Ordinaire	+.25
Verre Vert	+.14
Verre de Suède	+.56

and these are probably somewhat nearly correct, except the negative value for Choisy-le-Roi.

With the exception of Regnault, very few observers have taken up this subject. Among these, however, we may mention Recknagel, who has made the determination for common glass between 0° and 100°. I have found approximately the constants for my formula in this case, and have calculated the values in the fourth column of the following table.

---

* Comptes Rendus, lxix.

TABLE IV. — RECKNAGEL'S RESULTS COMPARED WITH THE FORMULA.

Air Thermometer.	Mercurial Thermometer.		Difference.
	Observed.	Calculated.	
0	0	0	0
10	10.08	10.08	0
20	20.14	20.14	0
30	30.18	30.18	0
40	40.20	40.20	0
50	50.20	50.20	0
60	60.18	60.18	0
70	70.14	70.15	+.01
80	80.10	80.11	+.01
90	90.05	90.06	+.01
100	100.00	0	0

$$b = 290° \qquad a = .000\,000\,33$$
$$T = t + at(100 - t)(b - t)$$

It will be seen that the values of the constants are not very different from those which satisfy Regnault's experiments.

There seems to be no doubt, from all the experiments we have now discussed, that the point of maximum difference is not at 50°, but at some less temperature, as 40° to 45°, and this agrees with my own experiments, and a recent statement by Ellis in the Philosophical Magazine. And I think the discussion has proved beyond doubt that the formula is sufficiently accurate to express the difference of the mercurial and air thermometers throughout at least a range of 200°, and hence is probably very accurate for the range of only 100° between 0° and 100°.

Hence it is only necessary to find the constants for my thermometers. But before doing this it will be well to see how exact the comparison must be. As the thermometers are to be used in a calorimetric research in which differences of temperature enter, the error of the mercurial compared with the air thermometer will be

$$\frac{dT}{dt} - 1 = a\left\{bt' - 2(b + t')t + 3t^2\right\},$$

which for the constants used in Recknagel's table becomes

$$\text{Error} = \frac{dT}{dt} - 1 = .000\,000\,33\left\{29000. - 780\,t + 3\,t^2\right\}.$$

This amounts to nearly one per cent at 0°, and thence decreases to 45°, after which it increases again. As only 0°.2 at the 40° point produces this large error at 0°, it follows that an error of only 0°.02 at 40° will produce an error of $\frac{1}{1000}$ at 0°. At other points the errors will be less.

Hence extreme care must be taken in the comparison and the most accurate apparatus must be constructed for the purpose.

2. Description of Apparatus.

The Air Thermometer.

In designing the apparatus, I have have had in view the production of a uniform temperature combined with ease of reading the thermometers, which must be totally immersed in the water. The uniformity, however, needed only to apply to the air thermometer and to the bulbs of the mercurial thermometer, as a slight variation in the temperature of the stems is of no consequence. A uniform temperature for the air thermometer is important, because it must take time for a mass of air to heat up to a given temperature within 0°.01 or less.

Fig. 1 gives a section of the apparatus. This consists of a large copper vessel, nickel-plated on the outside, with double walls an inch apart, and made in two parts, so that it could be put together watertight along the line ab. As seen from the dimensions, it required about 28 kilogrammes of water to fill it. Inside of this was the vessel $mdefghkln$, which could be separated along the line dk. In the upper part of this vessel, a piston, q, worked, and could draw the water from the vessel. The top was closed by a loose piece of metal, op, which fell down and acted as a valve. The bottom of this inner vessel had a false bottom, cl, above which was a row of large holes; above these was a perforated diaphragm, s. The bulb of the air thermometer was at t, with the bulbs of the mercurial thermometers almost touching it. The air thermometer bulb was very much elongated, being about 18$^{cm.}$ long and 3 to 5$^{cm.}$ in diameter. Although the bulbs of the thermometers were in the inner vessel, the stems were in the outer one, and the reading was accomplished through the thick glass window uv.

The change of the temperature was effected by means of a Bunsen burner under the vessel w.

The working of the apparatus was as follows. The temperature having been raised to the required point, the piston q was worked to stir up the water; this it did by drawing the water through the holes at cl and the perforated diaphragm s, and thence up through the apparatus to return on the outside. When the whole of the water is at a nearly uniform temperature the stirring is stopped, the valve $o\,p$ falls into place, and the connection of the water in the outer and inner vessels is practically closed as far as currents are concerned, and

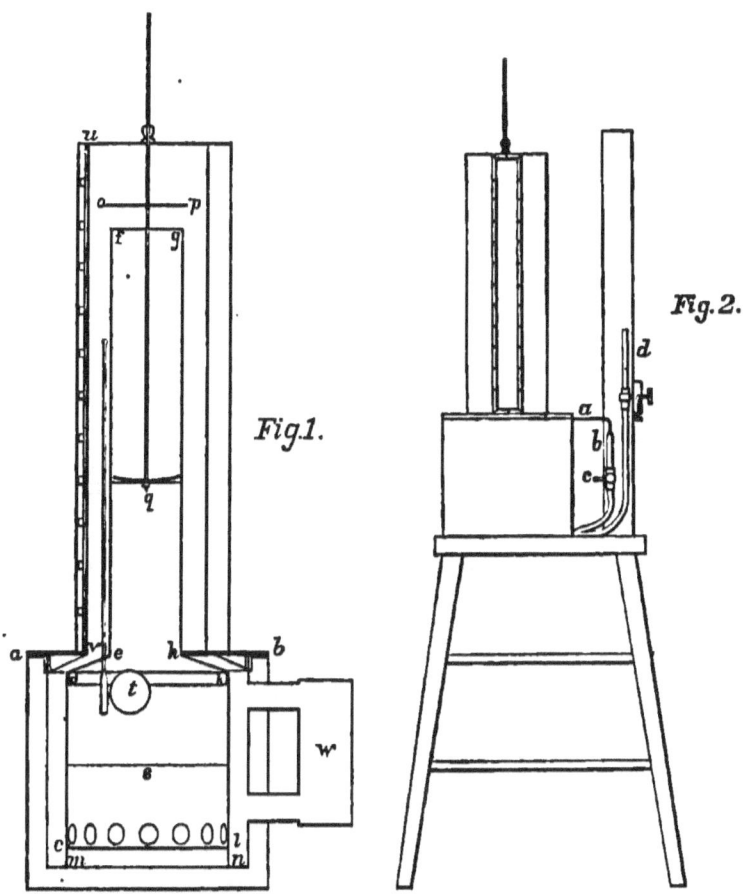

Fig.1.

Fig.2.

before the water inside can cool a little the outer water must have cooled considerably.

So effective was this arrangement that, although some of the thermometers read to $0°.007$ C., yet they would remain perfectly stationary for several minutes, even when at 40° C. At very high temperatures, such as 80° or 90° C., the burner was kept under the vessel w all the time, and supplied the loss of the outer vessel by radiation. The inner vessel would under these circumstances remain at a very constant temperature. The water in the outer vessel never differed by more than a small fraction of a degree from that in the inner one.

To get the 0 and 100° points the upper parts of the vessel above the line $a\,b$ were removed, and ice placed around the bulb of the air thermometer, and left for several hours, until no further lowering took place. For the 100° point the copper vessel shown in Fig. 3 was used. The portion y of this vessel fitted directly over the bulb of the air thermometer. On boiling water in x, the steam passed through the tube to the air thermometer. It is with considerable difficulty that the 100° point is accurately reached, and, unless care be taken, the bulb will be at a slightly lower temperature. Not only must the bulb be in the steam, but the walls of the cavity must also be at 100°. To accomplish this in this case, a large mass of cloth was heaped over the instrument, and then the water in x vigorously boiled for an hour or so. After fifteen minutes there was generally no perceptible increase of temperature, though an hour was allowed so as to make certain.

The external appearance of the apparatus is seen in Fig. 2. The method of measuring the pressure was in some respects similar to that used in the air thermometer of Jolly, except that the reading was taken by a cathetometer rather than by a scale on a mirror. The capillary stem of the air thermometer leaves the water vessel at a, and passes to the tube b, which is joined to the three-way cock c. The lower part of the cock is joined by a rubber tube to another glass tube at d, which can be raised and lowered to any extent, and has also a fine adjustment. These tubes were about 1.5$^{cm.}$ diameter on the inside, so that there should be little or no error from capillarity. Both tubes were exactly of the same size, and for a similar reason.

The three-way cock is used to fill the apparatus with dry air, and also to determine the capacity of the tube above a given mark. In filling the bulb, the air was pumped out about twenty times, and allowed to enter through tubes containing chloride of calcium, sulphuric acid, and caustic soda, so as to absorb the water and the carbonic acid.

The Cathetometer.

The cathetometer was one made by Meyerstein, and was selected because of the form of slide used. The support was round, and the telescope was attached to a sleeve which exactly fitted the support. The greatest error of cathetometers arises from the upright support not being exactly true, so that the telescope will not remain in level at all heights. It is true that the level should be constantly adjusted, but it is also true that an instrument can be made where such an adjustment is not necessary. And where time is an element in the accuracy, such an instrument should be used. In the present case it was absolutely necessary to read as quickly as possible, so as not to leave time for the column to change. In the first place the round column, when made, was turned in a lathe to nearly its final dimensions. The line joining the centres of the sections must then have been very accurately straight. In the subsequent fitting some slight irregularities must have been introduced, but they could not have been great with good workmanship.* The upright column was fixed, and the telescope moved around it by a sleeve on the other sleeve. Where

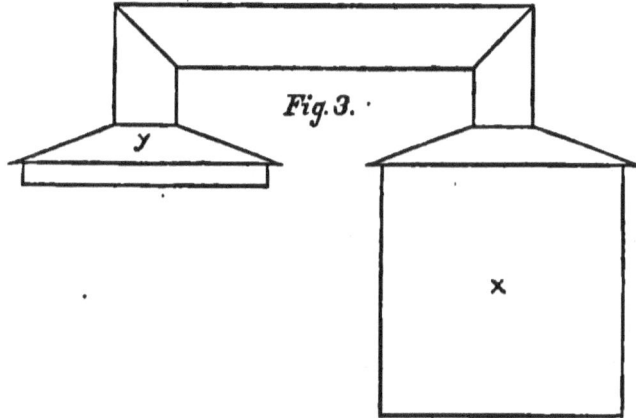

Fig. 3.

the objects to be measured are not situated at a very wide angle from each other, this is a good arrangement, and has the advantage that any side of the column can be turned toward the object, and so, even if it were crooked, we could yet turn it into such a position as to nearly eliminate error.

It was used at a distance of about 110$^{cm.}$ from the object, and no difficulty was found after practice in setting it on the column to $\frac{1}{50}$$^{mm.}$ at least. The cross hairs made an angle of 45° with the horizontal, as this was found to be the most sensitive arrangement.

The scale was carefully calibrated, and the relative errors † for the

---

* The change of level along the portion generally used did not amount to more than .1 of a division, or about .01$^{mm.}$ at the mercury column, as this is about the smallest quantity which could be observed on the level.

† These amounted to less than .016$^{mm.}$ at any part.

portion used were determined for every centimeter, the portion of the scale between the 0° and 100° points of the air thermometer being assumed correct. There is no object in determining the absolute value of the scale, but it should agree reasonably well with that on the barometer; for let H_0, H_t, and H_{100} be the readings of the barometer, and h_0, h_t, and h_{100} the readings of the cathetometer at the temperatures denoted by the subfix. Then approximately

$$t = \frac{(H_t + h_t) - (H_0 + h_0)}{(H_{100} + h_{100}) - (H_0 + h_0)} = \frac{H_t - H_0 + h_t - h_0}{H_{100} - H_0 + h_{100} - h_0}.$$

As the height of the barometer varies only very slightly during an experiment, the value of this expression is very nearly

$$\frac{h_t - h_0}{h_{100} - h_0},$$

which does not depend on the absolute value of the scale divisions.

But the best manner of testing a cathetometer is to take readings upon an accurate scale placed near the mercury columns to be measured. I tried this with my instrument, and found that it agreed with the scale to within two or three one-hundredths of a millimeter, which was as near as I could read on such an object.

In conclusion, every care was taken to eliminate the errors of this instrument, as the possibility of such errors was constantly present in my mind; and it is supposed that the instrumental errors did not amount to more than one or two one-hundredths of a millimeter on the mercury column. The proof of this will be shown in the results obtained.

The Barometer.

This was of the form designed by Fortin, and was made by James Green of New York. The tube was 2.0^{cm} diameter nearly on the outside, and about 1.7^{cm} on the inside. The correction for capillarity is therefore almost inappreciable, especially as, when it remains constant, it is exactly eliminated from the equation. The depression for this diameter is about $.08^{mm}$, but depends upon the height of the meniscus. The height of the meniscus was generally about 1.3^{mm}; but according as it was a rising or falling meniscus, it varied from 1.4 to 1.2^{mm}. These are the practical values of the variation, and would have been greater if the barometer had not been attached to the wall a little loosely, so as to have a slight motion when handled. Also in use the instrument was slightly tapped before reading. The variation of the height of the meniscus from 1.2 to 1.4$^{mm.}$ would affect the reading only to the extent of .01 to .02$^{mm.}$

The only case where any correction for capillarity is needed is in finding the temperatures of the steam at the 100° point, and will then affect that temperature only to the extent of about 0°.005.

The scale of the instrument was very nearly standard at 0° C., and was on brass.

At the centre of the brass tube which surrounded the barometer, a thermometer was fixed, the bulb being surrounded by brass, and therefore indicating the temperature of the brass tube.

In order that it should also indicate the temperature of the barometer, the whole tube and thermometer were wrapped in cloth until a thickness of about 5 or 6$^{cm.}$ was laid over the tube, a portion being displaced to read the thermometers. This wrapping of the barometer was very important, and only poor results were obtained before its use; and this is seen from the fact that 1° on the thermometer indicates a correction of .12$^{mm.}$ on the barometer, and hence makes a difference of 0°.04 on the air thermometer.

As this is one of the most important sources of error, I have now devised means of almost entirely eliminating it, and making continual reading of the barometer unnecessary. This I intend doing by an artificial atmosphere, consisting of a 'large vessel of air in ice, and attached to the open tube of the manometer of the air thermometer.

The Thermometers.

The standard thermometers used in my experiments are given in the following table.

TABLE V. — STANDARD THERMOMETERS.*

Mark.	Approximate Length in Centimeters.	Range.	Graduation.	Length occupied by 1° C.	Maker.	When made	Owner or Lender.
6167	65	0° to 100°	Millimeter.	... mm.	Baudin.	1876–7	Physical Laboratory, Johns Hopkins University.
6163	50	−6° to 40°	"	9.0	"	"	"
6165	50	−3° to 33°	"	11.7	"	"	"
6166	50	−2° to 31°	"	12.9	"	"	"
Kew 104	66	−21° to 108°	0°.5 F.	4.6	Welsh.	July, 1863	Prof. Barker, Univ. of Pennsylvania.
Geissler	45	−8° to 102°	0°.1 C.	3.8	Geissler.	(?)	Chemical Laboratory, Johns Hopkins University.
368	48	Nearly 0° to Nearly 100°	Arbitrary equal volumes	3.2	Fastré.	1851	Prof. Gibbs, Harvard Coll.
876	48	0° to 100°	"	3.1	"	"	"
3235	About 40	82° F. to 212° F. or more.	1° F.	1.6	Casella.	(?)	Prof. Pickering, Harvard Observatory.
32378	46	−15° to 110°	0°.2 C.	3.2	"	See Kew Register for 1873, −4, −8.	Prof. Trowbridge, Harvard College.
7316	61	−1° to +101°	0°.1 C.	4.9	Baudin.	Oct., 1878	Physical Laboratory, Johns Hopkins University.
7834	61	0° to 100°	0°.1 C.	4.9	"	"	S. W. Holman, Mass. Inst. of Technology.

* Besides these there were several other standards at my disposal. One from Salleron was not well calibrated, and several with millimeter scale, by Baudin, have not yet been calibrated. I have also several more by Baudin, graduated to 0°.2 C., which, although good, are not his best work.
The comparison of the last six, five of which were brought here by Mr. Holman for comparison, is given in the Appendix to this part of my paper. I am much indebted to Prof. Barker for the use of his standard, and to Mr. Holman for the use of his tables of comparison.

The calibration of the first four thermometers has been described. The calibration of the Kew standard was almost perfect, and no correction was thought necessary. The scale divided on the tube was to half-degrees Fahrenheit; but as the 32° and 212° points were not correct, it was in practice used as a thermometer with arbitrary divisions. The interval between the 0° and 100° points, as Welsh found it, was 180°.12, using barometer at 30 inches, or 180°.05 as corrected to 760$^{mm.}$ of mercury.* At the present time it is 179°.68,† showing a change of 1 part in 486 in twenty-five years. This fact shows that the ordinary method of correcting for change of zero is not correct, and that the coefficient of expansion of glass changes with time.‡

I have not been able to find any reference to the kind of glass used in this thermometer. But in a report by Mr. Welsh we find a comparison, made on March 19, 1852, of some of his thermometers with two other thermometers, — one by Fastré, examined and approved by Regnault, and the other by Troughton and Simms. The thermometer which I used was made a little more than a year after this; and it is reasonable to suppose that the glass was from the same source as the standards Nos. 4 and 14 there used. We also know that Regnault was consulted as to the methods, and that the apparatus for calibration was obtained under his direction.

I reproduce the table here with some alterations, the principal one of which is the correction of the Troughton and Simms thermometers, so as to read correctly at 32° and 212°, the calibration being assumed correct, but the divisions arbitrary.

---

* Boiling point, Welsh, Aug. 17, 1853, 212°.17; barometer 30$^{in.}$
Freezing point, " " " 32°.05.
Boiling point, Rowland, June 22, 1878, 212°.46; barometer 760^{mm}..
Freezing point, " " " 32°.78.
The freezing point was taken before the boiling point in either case.

† 179°.70, as determined again in January, 1879.

‡ The increase shown here is 1 in 80 nearly! It is evidently connected with the change of zero; for when glass has been heated to 100°, the mean coefficient of expansion between 0 and 100° often changes as much as 1 in 50. Hence it is not strange that it should change 1 in 80 in twenty-five years. I believe this fact has been noticed in the case of standards of length.

TABLE VI. — COMPARISON BY WELSH, 1852.

Mean of Kew Standards Nos. 4 and 14.	Fastré 231, Regnault.	Δ Kew.	Troughton and Simms (Royal Society).	Δ Kew.
°	°		°	
32.00	32.00	0	32.00	0
38.71	38.72	+.01	38.70	—.01
45.04	45.03	—.01	45.03	—.01
49.96	49.96	.00	49.96	.00
55.34	55.37	+.03	55.34	.00
60.07	60.05	—.02	60.06	—.01
65.39	65.41	+.02	65.36	—.03
69.93	69.95	+.02	60.93	.00
74.69	74.69	.00	74.72	+.03
80.05	80.06	+.01	80.14	+.09
85.30	85.33	+.03	85.44	+.14
90.50	90.51	+.01	90.56	+.06
95.26	95.24	—.02	95.40	+.14
101.77	101.77	.00	101.94	+.15
109.16	109.15	—.01	109.25	+.08
212.00	212.00	.00	212.00	.00

It is seen that the Kew standards and the Fastré agree perfectly, but that the Troughton and Simms standard stands above the Kew thermometers at 100° F.

The Geissler standard was made by Geissler of Bonn, and its scale was on a piece of milk glass, enclosed in a tube with the stem. The calibration was fair, the greatest error being about 0°.015 C., at 50° C.; but no correction for calibration was made, as the instrument was only used as a check for the other thermometers.

3. RESULTS OF COMPARISON.

Calculation of Air Thermometer.

This has already been described, and it only remains to discuss the formula and constants, and the accuracy with which the different quantities must be known.

The well-known formula for the air thermometer is

$$T = \frac{1}{a} \frac{H - h + \frac{v}{V}\left\{ H \frac{1+bt'}{1+at'} - h \frac{1+bt}{1+at} \right\} + bHT}{h - \frac{v}{V}\left\{ H \frac{1+bt'}{1+at'} - h \frac{1+bt}{1+at} \right\}}.$$

Solving with reference to T, and placing in a more convenient form, we have

$$T = \frac{1}{a} \frac{H - h' + \frac{v}{V} H \frac{1}{1+\gamma t}}{h' - \frac{v}{V} H \frac{1}{1+\gamma t} - H \frac{b}{a}} \text{ nearly,}$$

where
$$h' = h\left(1 + \frac{v}{V}\frac{1}{1+\gamma t'}\right),$$

and
$$\gamma = a - b = .00364.$$

For the first bulb, $\frac{v}{V} = .0057.$

For the second bulb, $\frac{v}{V} = .0058.$

To discuss the error of T due to errors in the constants, we must replace a by its experimental value, seeing that it was determined with the same apparatus as that by which T was found. As it does not change very much, we may write approximately

$$T = 100\,\frac{H-h}{H_{100}-h}\left\{1 - \frac{v}{V}\frac{1}{1+\gamma t}\left(\frac{H_{100}-H}{h}\right) - \frac{b_{100}H_{100}-bH}{ah}\right\}.$$

From this formula we can obtain by differentiation the error in each of the quantities, which would make an error of one tenth of one per cent in T. The values are for $T = 40°$ nearly; $t = 20°$; $H_{100} - h = 270^{mm}$; and $h = 750^{mm}$. If x is the variable,

$$\Delta x = \frac{dx}{dT}\,\Delta T = \frac{dx}{dT}\frac{T}{1000} = .04\,\frac{dx}{dT}$$

TABLE VII. — Errors producing an Error in T of 1 in 1000 at 40° C.

	H	H_{100} or h.	$\frac{v}{V}$	$\frac{b_{100}}{a}$, $\frac{b}{a}$ constant.	$\frac{b}{a}$, $\frac{b_{100}}{a}$ constant.	$\frac{b_{100}}{a}$, $\frac{b_{100}-b}{a}$ const.	$\frac{b_{100}-b}{a}$, $\frac{b_{100}}{a}$ constant.
Absolute value, Δx	$.11^{mm}$	$.27^{mm}$	.005	.00074	.00087	.0047	.00087
Relative value, $\frac{\Delta x}{x}$	...	...	0.9	.10	.12	.62	...

From this table it would seem that there should be no difficulty in determining the 40° point on the air thermometer to at least 1 in 2000; and experience has justified this result. The principal difficulty is in the determination of H, seeing that this includes errors in reading the barometer as well as the cathetometer. For this reason, as mentioned before, I have designed another instrument for future use, in which the barometer is nearly dispensed with by use of an artificial atmosphere of constant pressure.

The value of $\frac{v}{V}$ does not seem to affect the result to any great extent; and if it was omitted altogether, the error would be only about 1 in 1,000, assuming that the temperature t was the same at the determination of the zero point, the 40° point, and the 100° point. It seldom varied much.

The coefficient of expansion of the glass influences the result very slightly, especially if we know the *difference* of the mean coefficients between 0° and 100°, and say —10° and +10°. This difference I at first determined from Regnault's tables, but afterwards made a determination of it, and have applied the correction.*

The table given by Regnault is for one specimen of glass only; and I sought to better it by taking the expansion at 100° from the mean of the five specimens given by Regnault on p. 231 of the first volume of his *Relation des Expériences*, and reducing the numbers on page 237 in the same proportion. I thus found the values given in the second column of the following table.

TABLE VIII. — COEFFICIENT OF EXPANSION OF THE GLASS OF THE AIR THERMOMETER, ACCORDING TO THE AIR THERMOMETER.

Temperature according to Air Thermometer.	Values of b used for a first Calculation	b from Regnault's Table, Glass No. 5.	Experimental Results.			
			Apparent Coefficient of Expansion of Mercury.	b, using Regnault's Value for Mercury.†	b, using Recknagel's Value for Mercury.‡	b, using Wüllner's Value for Mercury.§
0	.0000252	.0000263	.....	.....	.....	.....
20°	.0000253	.0000264	.00015410	.0000254	.0000264	.0000273
40°	.0000256	.0000267	.00015395	.0000258	.0000266	.0000276
60°	.0000259	.0000270	.00015391	.0000261	.0000267	.0000278
80°	.0000262	.0000273	.....	.....	.....	.....
100°	.0000264	.0000276	.00015381	.0000277	.0000277	.0000287

The second column contains the values which I have used, and one of the last three columns contains my experimental results, the last being probably the best. The errors by the use of the second column compared with the last are as follows: —

$\frac{1}{10000}$ from using $b_{100} - b_{40} = .0000008$ instead of .0000011;

$\frac{1}{7000}$ from using $b_{100} \phantom{- b_{40}} = .0000264$ instead of .0000287;

or, $\frac{1}{1000}$ for both together.

---

* This was determined by means of a large weight thermometer in which the mercury had been carefully boiled. The glass was from the same tube as that of the air thermometer, and they were cut from it within a few inches of each other.

† Relation des Expériences, i. 328.

‡ Pogg. Ann., cxxiii. 135. § Experimental Physik, Wüllner, i. 67.

As the error is so small, I have not thought it worth while to entirely recalculate the tables, but have calculated a table of corrections as follows, and have so corrected them: —

TABLE IX. — TABLE OF CORRECTIONS.

T' Calculated Temperature.	T Corrected Temperature.	Correction.
0°	0°	0
10	9.9971	—.0029
20	19.9946	—.0054
30	29.9924	—.0076
40	39.9907	—.0093
50	49.9894	—.0106
60	59.9865	—.0135
80	79.9880	—.0120
100	100.	0

$$T = T' \{1 + 373 (b'_{100} - b_{100}) - (273 + T) (b' - b)\}.$$

$$T = T' \{1 - .000858 + (273 + T') (b - b')\}.$$

$T = .99975\ T'$ approximately between 0 and 40°. This last is true within less than $\frac{1}{1000}$ of a degree.

The two bulbs of the air thermometer used were from the same piece of glass tubing, and consequently had nearly, if not quite, the same coefficient of expansion.

In the reduction of the barometer and other mercurial columns to zero, the coefficient .000162 was used, seeing that all the scales were of brass.

In the tables the readings of the thermometers are reduced to volumes of the tube from the tables of calibration, and they are corrected for the pressure of water, which increased their reading, except at 0°, by about 0°.01 C.

The order of the readings was as follows in each observation: — 1st, barometer; 2d, cathetometer; 3d, thermometers forward and backward; 4th, cathetometer; 5th, barometer, &c., — repeating the same once or twice at each temperature. In the later observations, two series like the above were taken, and the water stirred between them.

The following results were obtained at various times for the value of a with the first bulb: —

.0036664
.0036670
.0036658
.0036664
.0036676

Mean $a =$ .00366664

obtained by using the coefficient of expansion of glass .0000264 at 100°, or $a =$ .0036698, using the coefficient .0000287.

The thermometers Nos. 6163, 6165, 6166, were always taken out of the bath when the temperature of 40° was reached, except on November 14, when they remained in throughout the whole experiment.

The thermometer readings are reduced to volumes by the tables of calibration.

TABLE X. — 1st Series, Nov. 14, 1877.

Relative Weight.	Air Thermometer.	V 6163.	V 6166.	V 6167.	Temperature by 6167.	Δ
4	0	115.33	21.25	6.147	0	0
4	17°.1425	422.84	255.80	15.685	17°.661	.286
4	23°.793	534.71	341.05	19.157	24°.089	.296
5	30°.582	653.49	431.71	22.833	30°.896	.314
2	38°.569	798.18	...	27.175	38°.935	.366
2	51°.040	...	...	33.864	51°.320	.280
4	59°.137	...	...	38.256	59°.452	.315

The first four series, Tables X. to XIII., were made with one bulb to the air thermometer. A new bulb was now made, whose capacity was 192.0$^{c.cm.}$, that of the old being 201.98$^{c.cm.}$. The value of $\frac{v}{V}$ for the new bulb was .0058. The values of h' and a were obtained as follows: —

	a	h'
June 8th	.00366790	753.876
June 22d	.00366977	753.805
June 25th	.00366779	753.837
Mean	.0036685	753.84

This value of a is calculated with the old coefficient for glass. The new would have given .0036717.

TABLE XI. — Second Series, November 20-21, 1877.

Relative Weight.	Barometer corrected to 0° C.	Difference of Columns reduced to 0°	H	H — h'	$\frac{V}{v}\frac{H}{1+\gamma t}$	$H\frac{b}{a}$	T Air Thermometer.	V 6168.	V 6166.	V 6167.	Geissler.	Temperature by 6167.	Temperature by Geissler.
2	.....	—27.28	.....	.....	.....	.....	0	114.94	21.09	6.128	+0.50	0	0
3	773.00	— 6.84	766.16	16.49	4.09	5.23	7°.581	248.66	123.10	10.282	8.121	7°.689	7°.57
2	772.78	+16.72	789.50	39.83	4.22	5.43	16°.234	400.52	238.60	14.986	16.78	16°.395	10°.38
2	772.32	48.27	820.59	70.92	4.35	5.68	27°.755	602.80	392.65	21.246	28.31	27°.982	27°.98
2	772.21	50.25	822.46	72.79	4.36	5.69	28°.449	615.10	402.29	21.628	29.01	28°.689	28°.67
1	772.27	84.17	856.44	106.77	4.54	5.97	41°.070	834.99	.....	28.439	41.53	41°.295	41°.26
8	772.34	88.05	856.39	106.72	4.53	5.96	40°.678	828.87	.....	28.259	41.23	40°.962	40°.96
2	772.61	104.08	876.69	127.02	4.68	6.14	48°.614	.....	.....	32.516	49.01	48°.841	48°.79
2	772.25	112.78	885.08	135.86	4.70	6.21	51°.708	.....	.....	34.208	52.17	51°.973	51°.96
4	771.93	146.68	918.61	168.94	4.87	6.50	64°.206	.....	.....	40.943	64.47	64°.439	64°.34
3	771.64	171.67	943.31	198.64	4.98	6.71	73°.402	.....	.....	45.800	73.61	73°.560	73°.53
3	771.24	201.42	972.66	222.99	5.12	6.95	84°.344	.....	.....	51.767	84.43	84°.473	84°.41
3	771.07	201.59	972.66	222.99	5.12	6.95	84°.344	.....	.....	51.770	84.43	84°.479	84°.41
0	770.90	243.10	1014.00	264.33	5.32	7.31	99°.781	.....	.....	60.064	.....	99°.830	.....
0	770.71	242.12	1012.88	263.16	5.82	7.80	99°.846	.....	.....	.....	99.26	99°.350	99°.38
...	.....	.....	.....	.....	.....	.....	100°.000	.....	.....	60.166	99.93	.....	.....

$h' = 749.67$ mm.

TABLE XII. — THIRD SERIES, JANUARY 25, 1878.

Relative Weight.	Barometer reduced to 0°	Difference of Columns reduced to 0°	H	$H - h'$	$\frac{v}{V}\frac{H}{1+\gamma t}$	$H\frac{b}{a}$	Air Thermometer.	V 6168.	V 6166.	Geissler.	Temperature by Geissler.
1	764.193	−17.644	746.549	.....	3.94	...	0	115.50	21.44	+0.60	0
1	768.00	− 2.69	760.40	9.91	3.99	5.19	5°.114	205.80	90.84	.....	.....
1	763.06	− 0.63	762.43	11.94	4.00	5.21	5°.865	218.60	99.87	.....	.....
1	762.87	+11.96	774.83	24.34	4.06	5.31	10°.451	299.94	162.17	11.12	10°.58
1	762.82	34.09	796.91	46.42	4.18	5.49	18°.628	448.14	271.18	19.30	18°.82
1	762.89	54.47	817.36	66.87	4.29	5.65	26°.207	576.52	372.70	26.91	26°.48
1	762.89	62.19	825.08	74.50	4.30	5.72	29°.057	626.57	410.98	29.73	29°.31
1	762.88	88.27	851.10	100.61	4.43	5.98	38°.700	794.08	.....	39.32	38°.96
1	762.82	87.16	849.98	99.49	4.42	5.92	38°.288	787.86	.....	38.05	38°.59

$N = 750.49.$

TABLE XIII.—FOURTH SERIES, FEBRUARY 11, 1878.

Relative Weight.	Barometer reduced to 0°	Difference of Columns reduced to 0°	H	$H \div h'$	$\frac{v}{V} \frac{H}{1+\gamma t}$	$H \frac{b}{a}$	T Air Ther- mometer.	V 6164.	V 6165.	V 6163.	Geissler.	Tempera- ture by Geissler.
3	755.810	—8.737	746.573	—3.986	4.007	.....	0°.008	21.21	6.202	115.60	+0.60	0
3	755.424	+0.611	756.035	5.447	4.039	5.156	8°.489	67.72	13.736	176.30	.....	.....
2	755.846	1.600	756.946	6.358	4.042	5.162	8°.825	71.08	14.561	182.78	.....	.....
3	755.274	11.665	766.939	16.351	4.095	5.239	7°.521	122.20	22.614	247.85	.....	.....
3	755.250	12.792	768.042	17.454	4.097	5.246	7°.027	127.29	23.424	254.38	8.56	8°.01
4	755.222	25.536	780.758	30.170	4.169	5.357	12°.634	190.71	33.793	337.90	13.81	12°.79
4	755.495	40.863	796.358	45.770	4.241	5.479	18°.406	268.20	46.404	489.43	19.11	18°.63
4	755.444	58.882	814.326	63.738	4.336	5.599	25°.060	356.86	60.841	555.71	25.72	25°.28
3	755.481	71.852	827.333	76.745	4.300	5.733	29°.876	421.23	71.294	640.19	30.55	30°.14
3	755.497	71.365	826.862	76.274	4.386	5.735	29°.700	418.61	70.881	636.79	30.35	29°.94
4	755.923	94.718	850.641	100.053	4.519	5.029	38°.522	.....	.....	791.26	30.15	38°.80
5	755.626	258.758	1015.384	264.796	5.397	7.321	99°.840	.....	.....	.....	99.813	99°.84
3	755.360	—8.743	746.617	—4.029	—4.007	.....	—0°.008	.....	.....	.....	.....	.....

$h' = 760.588$ mean value before and after experiments.

The value $\alpha = .00\,366\,707$ as obtained on the same day was used in this calculation.

TABLE XIV. — FIFTH SERIES, JUNE 8, 1878.

Weight for Air Thermometer.	Barometer reduced to 0°	Difference of Columns reduced to 0°	H	$H - h'$	$\frac{v}{V} \cdot \frac{H}{1+\gamma t}$	$\frac{b}{H} - \frac{1}{a}$	T Air Thermometer.	V 6165.	V 6163.	Kew Standard No. 104.	V 6168. Shortened Column.	Geissler.	Temperature by Kew Standard.	Temperature by Geissler.	Δ Kew.	Δ Geissler.
2	750.819	−1.004	749.815	−4.025	4.061	5.281	0°.014	7.054	115.93	32.78	......	+0.60	0	0	−.01	−.01
2	749.710	21.904	771.614	17.774	4.164	5.343	8°.035	24.589	257.14	47.23	......	......	8°.04	......	+.01	......
2	749.612	30.625	780.237	26.397	4.208	5.420	11°.209	31.445	312.58	52.91	62.86	......	11°.20	......	−.01	......
1	740.281	39.880	789.161	35.321	4.253	5.483	14°.496	38.594	370.22	58.84	106.67	15.155	14°.50	14°.65	.00	+.15
2	748.986	48.089	797.075	43.235	4.298	5.546	17°.412	46.006	421.92	64.09	146.16	18.12	17°.43	17°.63	+.02	+.22
2	747.814	57.014	804.828	50.988	4.332	5.680	20°.270	51.277	472.82	69.82	184.54	21.00	20°.34	20°.53	+.07	+.26
2	747.530	74.320	821.850	68.010	4.418	5.785	25°.546	64.930	582.14	80.55	258.04	27.25	25°.50	25°.81	+.04	+.28
2	747.422	87.619	835.041	81.201	4.465	5.966	31°.412	76.485	667.40	89.29	333.17	32.14	31°.45	31°.73	+.04	+.32
1	747.082	108.744	855.826	101.986	4.596	6.211	39°.088	......	801.76	103.06	435.58	39.80	39°.11	39°.44	+.02	+.35
2	747.008	139.450	886.458	132.618	4.754	6.423	50°.407	......	......	123.44	......	50.99	50°.46	50°.70	+.05	+.29
2	746.816	164.446	911.262	157.422	4.890	7.346	59°.587	......	......	139.87	......	59.99	59°.60	59°.75	+.01	+.16
2	746.826	272.075	1018.901	265.061	5.463	......	99°.518	......	......	......	......	......	......	......	......	......

TABLE XV. — SIXTH SERIES, JUNE 22, 1878.

Relative Weight.	Barometer reduced to 0°.	Difference of Columns reduced to 0°.	H	H − h'	$\frac{H}{1+\gamma t}$	$H\frac{b}{a}$	T Air Thermometer.	V 6165.	V 6163.	Kew No. 104.	V 6168.	Geissler.	Temperature by Kew Standard.	Temperature by Geissler.	Δ Kew.	Δ Geissler.
1	750.411	−0.628	749.783	−4.057	4.020	.....	−0°.010	7.040	115.83	32.75	21.63	+0.60	0	0	+.01	+.01
1	750.843	45.692	796.035	42.195	4.269	5.476	17°.021	44.186	415.33	63.48	249.81	17.71	17°.10	17°.22	+.08	+.20
1	750.220	82.935	833.155	79.315	4.448	5.774	50°.705	74.011	655.65	88.08	433.02	31.43	30°.79	31°.02	+.09	+.31
1	750.170	108.407	858.577	104.737	4.575	5.986	40°.086	.....	819.68	104.93	.....	40.79	40°.16	40°.43	+.07	+.84
1	749.914	136.270	886.184	132.344	4.721	6.212	50°.292	.....	.....	123.26	.....	50.87	50°.36	50°.57	+.07	+.28
1	750.127	173.486	923.613	169.778	4.919	6.531	64°.144	.....	.....	148.105	.....	64.515	64°.19	64°.30	+.05	+.17
1	749.994	190.045	940.039	186.199	5.004	6.665	70°.227	.....	.....	159.03	.....	70.555	70°.27	70°.37	+.04	+.14
1	750.653	212.443	963.096	209.256	5.127	6.867	78°.775	.....	.....	174.31	.....	78.97	78°.77	78°.84	+.00	+.07
1	750.923	231.446	982.369	228.529	5.220	7.024	85°.924	.....	.....	187.03	.....	86.01	86°.85	85°.93	−.07	+.01
1	751.304	268.164	1019.468	265.618	5.425	7.351	99°.700	.....	.....	211.88	.....	99.68	99°.68	99°.68	−.02	−.02

It now remains to determine from these experiments the most probable values of the constants in the formula, comparing the air with the mercurial thermometer. The formula is, as we have found,

$$t = T - at(t' - t)(b - t);$$

but I have generally used it in the following form: —

$$t = CV - t_0 - mt(100 - t)(1 - n(100 + t)),$$
$$t = C'V - t'_0 - mt(40 - t)(1 - n(40 + t)).$$

And the following relations hold among the constants: —

$$C = C'(1 + m(60 - 8400 n)), \text{ nearly,}$$
$$a = mn,$$
$$b = \frac{1}{n} - 100°,$$
$$T = CV - t_0,$$
$$t_0 = t'_0 \frac{C}{C'}.$$

In these formulæ t is the temperature on the air thermometer; V is the volume of the stem of the mercurial thermometer, as determined from the calibration and measured from any arbitrary point; and C', t'_0, m, and n are constants to be determined.

The best way of finding these is by the method of least squares. C' must be found very exactly; t_0 is only to be eliminated from the equations; m must be found within say ten per cent, and n need only be determined roughly. To find them only within these limits is a very difficult matter.

Determination of n.

As this constant needs a wide range of temperatures to produce much effect, it can only be determined from thermometer No. 6167, which was of the same glass as 6163, 6165, and 6166. It is unfortunate that it was broken on November 21, and so we only have the experiments of the first and second series. From these I have found $n = .003$ nearly. This makes $b = 233°$, which is not very far from the values found before from experiments above 100° by Regnault on ordinary glass.*

---

* Some experiments with Baudin thermometers at high temperatures have given me about 240°, — a remarkable agreement, as the point must be uncertain to 10° or more.

Determination of C and m.

I shall first discuss the determination of these for thermometers Nos. 6163, 6165, and 6166, as these were the principal ones used.

As No. 6163 extended from $0°$ to $40°$, and the others only from 0 to $30°$, it was thought best to determine the constants for this one first, and then find those for 6165 and 6166 by comparison. As this comparison is deduced from the same experiments as those from which we determine the constants of 6163, very nearly the same result is found as if we obtained the constants directly by comparison with the air thermometer.

The constants of 6163 can be found either by comparison with 6167, or by direct comparison with the air thermometer. I shall first determine the constants for No. 6167.

The constants C and t_0 for this thermometer were found directly by observation of the $0°$ and $100°$ points; and we might assume these, and so seek only for m. In other words, we might seek only to express the difference of the thermometers from the air thermometer by a formula. But this is evidently incorrect, seeing that we thus give an infinite weight to the observations at the 0 and $100°$ points. The true way is obviously to form an equation for each temperature, giving each its proper weight. Thus from the first series we find for No. 6167, —

Weight.	Equations of condition.		
4	0	$= 6.147\ C - t_0$,	
4	$17°.427$	$= 15.685\ C - t_0$	$- 930\ m$,
4	$23°.793$	$= 19.157\ C - t_0$	$- 1140\ m$,
&c.	&c.	&c.	
5	$100°$	$= 60.156\ C - t_0$,	

which can be solved by the method of least squares. As t_0 is unimportant, we simply eliminate it from the equations. I have thus found, —

Weight.			
1	Nov. 14	$C = 1.85171$	$m = .000217$
2	Nov. 20, 21	$C = 1.85127$	$m = .000172$
	Mean	$C = 1.85142$	$m = .000187$

The difference in the values of m is due to the observations not being so good as were afterwards obtained. However, the difference only signifies about $0°.03$ difference from the mean at the $50°$ point. After November 20 the errors are seldom half of this, on account of the greater experience gained in observation.

The ratio of C for 6167 and 6163 is found in the same way.

Weight.		
1	Nov. 14	.0310091
2	Nov. 20	.0309846
	Mean	.0309928

Hence for 6163 we have in this way

$$C = .057381 \qquad C' = .056995 \qquad m = .000187.$$

By direct comparison of No. 6163 with the air thermometer, we find the following.

Date.	Weight	C'.	m.
Nov. 14	1	.056920	.000239
Nov. 20	2	.056985	.000166
Jan. 25	3	.056986	.000226
Feb. 11	4	.056997	.000155
June 8	3	.056961	.000071
June 22	2	.056959	.000115
	Mean	$.056976 \pm .000004$	$.000154 \pm .000010$

The values of C' agree with each other with great exactness, and the probable error is only $\pm 0.°003$ C. at the 40° point.

The great differences in the values of m, when we estimate exactly what they mean in degrees, also show great exactness in the experiments. The mean value of m indicates a difference of only $0°.05$ between the mercurial and air thermometer at the 20° point, the 0° and 40° points coinciding. The probable error of m in degrees is only $\pm 0°.003$ C.

There is one more method of finding m from these experiments; and that is by comparing the values of C' with No. 6167, the glass of 6167 being supposed to be the same as that of 6163.

We have the formula

$$C = C' (1 + 34.8\ m).$$

Hence
$$m = \frac{C - C'}{34.8\ C'}.$$

We thus obtain the following results: —

Date.	Weight.	Value of m.
Nov. 14	1	.000236
Nov. 20	2	.000218
Jan. 25	3	.000217
Feb. 11	4	.000197
June 8	3	.000215
June 22	2	.000216
	Mean	.000213

The results for m are then as follows:—

From direct comparison of No. 6167 with the air thermometer .000187
From   "   "   " No. 6163   "   "   ·   "   .000154
From comparison of No. 6163 with No. 6167   .000213

The first and last are undoubtedly the most exact numerically, but they apply to No. 6167, and are also, especially the first, derived from somewhat higher temperatures than the 20° point, where the correction is the most important. The value of m, as determined in either of these ways, depends upon the determination of a difference of temperature amounting to 0°.30, and hence should be quite exact.

The value of m, as obtained from the direct comparison of No. 6163 with the air thermometer, depends upon the determination of a difference of about 0°.05 between the mercurial and the air thermometer. At the same time, the comparison is direct, the temperatures are the same as we wish to use, and the glass is the same. I have combined the results as follows:—

m from No. 6167   .000200
m "   6163   .000154
          Mean   .00018*

It now remains to deduce from the tables the ratios of the constants for the different thermometers.

The proper method of forming the equations of condition are as follows, applying the method to the first series:—

Weight.
4    $21.25\ C_{III} = 115.33\ C_I - v_0$
4    $255.80\ C_{III} = 422.84\ C_I - v_0$
4    $341.05\ C_{III} = 534.71\ C_I - v_0$
5    $431.71\ C_{III} = 653.49\ C_I - v_0$

where C_{III} is the constant for No. 6166, C_I is that for No. 6163, and v_0 is a constant to be eliminated. Dividing by C_I, the equations can be solved for $\frac{C_{III}}{C_I}$. The following table gives the results.

---

* See Appendix to Thermometry, where it is finally thought best to reject the value from No. 6167 altogether.

TABLE XVI. — Ratios of Constants.

Date.	Weight.	$\frac{6163}{6167}$	$\frac{6166}{6167}$	$\frac{6166}{6163}$	$\frac{6165}{6163}$	$\frac{6165}{6166}$
Nov. 14	1	.031009	.040658	1.3111	....	.....
Nov. 20	2	.030985	.040670	1.3128	....	.....
Jan. 25	3	....	....	1.3122	....	.....
Feb. 11	4	....	....	1.3115	8.0588	6.1449
June 8	3	....	....	1.3108	8.0605	6.1469
June 22	2	....	....	1.3122	8.0588	6.1428
Mean		.030993 ±.00005	.040666 ±.000003	1.31175 ± .0004	8.0594 ± .0002	6.1451 ± .0004

From these we have the following, as the final most probable results : —

$$C_{ii} = 8.0601\ C_i,$$
$$C_{iii} = 1.31175\ C_i,$$
$$C_i = .031003\ C_{iv},$$
$$C_{ii} = .24991\ C_{iv},$$
$$C_{iii} = .040661\ C_{iv},$$

of which the last three are only used to calculate the temperatures on the mercurial thermometer, and hence are of little importance in the remainder of this paper.

The value of C' which we have found for the old value of the coefficient of expansion of glass was

$$C' = .056976;$$

and hence, corrected to the new coefficient, it is, as I have shown,

$$C_i = .056962.$$

Hence, $\quad C_{ii} = .45912,$
$\quad\quad\quad C_{iii} = .074720.$

And we have finally the three following equations to reduce the thermometers to temperatures on the air thermometer : —

Thermometer No. 6163:

$$T = .056962\ V' - t_0' - .00018\ T\ (40 - T)\ (1 - .003\ (T + 40)).$$

Thermometer No. 6165:

$$T = .45912\ V'' - t_0'' - .00018\ T\ (T - 40)\ (1 - .003\ (T + 40)).$$

Thermometer No. 6166:

$$T = .074720\, V''' - t_0''' - .00018\, T(T-40)\left(1 - .003\,(T+40)\right);$$

where V', V'', and V''' are the volumes of the tube obtained by calibration; t_0', t_0'', and t_0''' are constants depending on the zero point, and of little importance where a difference of temperature is to be measured; and T is the temperature on the air thermometer.

On the mercurial thermometer, using the 0° and 100° points as fixed, we have the following by comparison with No. 6167: —

Thermometer No. 6163; $t = .057400\ V - t_0$;
Thermometer No. 6165; $t = .46265\ V - t_0$;
Thermometer No. 6166; $t = .075281\ V - t_0$.

The Kew Standard.

The Kew standard must be treated separately from the above, as the glass is not the same. This thermometer has been treated as if its scale was arbitrary.

In order to have variety, I have merely plotted all the results with this thermometer, including those given in the Appendix, and drawn a curve through them. Owing to the thermometer being only divided to ½° F., the readings could not be taken with great accuracy, and so the results are not very accordant; but I have done the best I could, and the result probably represents the correction to at least 0°.02 or 0°.03 at every point.

(d) Reduction to the Absolute Scale.

The correction to the air thermometer to reduce to the absolute scale has been given by Joule and Thomson, in the Philosophical Transactions for 1854; but as the formula there used is not correct, I have recalculated a table from the new formula used by them in their paper of 1862.

That equation, which originated with Rankine, can be placed in the form

$$\frac{p\,v}{\mu} = C\left(1 - m\,\frac{\mu_0}{\mu^2}\, D\right);$$

where p, v, and μ are the pressure, volume, and absolute temperature of a given weight of the air; D is its density referred to air at 0° C. and 760^{mm} pressure; μ_0 is the absolute temperature of the freezing point; and m is a constant which for air is 0°.33 C.

For the air thermometer with constant volume

$$T = 100 \frac{p_t - p_0}{p_{100} - p_0};$$

$$\therefore T = (\mu - \mu_0)\left(1 + m\,D\left(\frac{1}{\mu} - \frac{1}{\mu_{100}}\right)\right);$$

or, since $D = 1$,

$$\mu - \mu_0 = T - .00088\,T\,\frac{100 - T}{273 + T},$$

from which I have calculated the following table of corrections.

TABLE XVII. — REDUCTION OF AIR THERMOMETER TO ABSOLUTE SCALE.

T Air Thermometer.	$\mu - \mu_0$ Absolute Temperature.	Δ or Correction to Air Thermometer.
0	0	0
10	9.9972	—.0028
20	19.9952	—.0048
30	29.9939	—.0061
40	39.9933	—.0067
50	49.9932	—.0068
60	59.9937	—.0063
70	69.9946	—.0054
80	79.9956	—.0044
90	89.9978	—.0022
100	100.000	0
200	200.037	+.037
300	300.092	+.092
400	400.157	+.157
500	500.228	+.228

It is a curious circumstance, that the point of maximum difference occurs at about the same point as in the comparison of the mercurial and air thermometers.

From the previous formula, and from this table of corrections, the following tables were constructed.

TABLE XVIII.—Thermometer No. 6163.

Reading in Millimeters on Stem.	Temperature on Mercurial Thermometer, 0 and 100° fixed.	Temperature on Mercurial Thermometer, 0 and 40° fixed by Air Thermometer.	Temperature on Air Thermometer.	Temperature on Absolute Scale from 0° C.	Reading in Millimeters on Stem.	Temperature on Mercurial Thermometer, 0 and 100° fixed.	Temperature on Mercurial Thermometer, 0 and 40° fixed by Air Thermometer.	Temperature on Air Thermometer.	Temperature on Absolute Scale from 0° C.
50	−.923°	−.917°	−.911°	−.911°	240	20.557°	20.409°	20.350°	20.345°
58.1	0	0	0	0	250	21.670	21.515	21.457	21.452
60	+.217	+.215	+.214	+.214	260	22.776	22.616	22.559	22.554
70	1.356	1.336	1.328	1.328	270	23.884	23.713	23.657	23.652
80	2.494	2.475	2.461	2.460	280	24.989	24.810	24.755	24.750
90	3.631	3.604	3.584	3.583	290	26.093	25.907	25.854	25.848
100	4.767	4.733	4.707	4.706	300	27.200	27.006	26.956	26.950
110	5.903	5.860	5.829	5.827	310	28.311	28.108	28.060	28.056
120	7.036	6.986	6.950	6.948	320	29.425	29.214	29.169	29.163
130	8.170	8.111	8.071	8.069	330	30.541	30.324	30.282	30.276
140	9.304	9.237	9.193	9.190	340	31.662	31.436	31.398	31.392
150	10.436	10.361	10.314	10.311	350	32.782	32.548	32.514	32.508
160	11.568	11.485	11.435	11.432	360	33.903	33.660	33.630	33.624
170	12.700	12.608	12.556	12.553	370	35.023	34.773	34.748	34.742
180	13.829	13.730	13.676	13.672	380	36.143	35.884	35.864	35.857
190	14.957	14.850	14.794	14.790	390	37.261	36.994	36.979	36.972
200	16.081	15.966	15.909	15.905	400	38.377	38.103	38.094	38.087
210	17.203	17.080	17.022	17.018	410	39.492	39.210	39.206	39.199
220	18.322	18.191	18.132	18.127	420	40.604	40.314	40.316	40.309
230	19.440	19.301	19.242	19.237					

TABLE XIX.—Thermometer No. 6165.

Reading in Millimeters on Stem.	Temperature on Mercurial Thermometer, 0 and 100° fixed.	Temperature on Mercurial Thermometer, 0 and 40° fixed by Air Thermometer.	Temperature on Air Thermometer.	Temperature on Absolute Scale from 0° C.	Reading in Millimeters on Stem.	Temperature on Mercurial Thermometer, 0 and 100° fixed.	Temperature on Mercurial Thermometer, 0 and 40° fixed by Air Thermometer.	Temperature on Air Thermometer.	Temperature on Absolute Scale from 0° C.
30	−.464°	−.460°	−.457°	−.457°	230	17.198°	17.067°	17.009°	17.005°
35	0	0	0	0	240	18.056	17.920	17.861	17.857
40	+.463	+.460	+.457	+.457	250	18.917	18.773	18.714	18.709
50	1.387	1.376	1.368	1.368	260	19.771	19.621	19.562	19.557
60	2.307	2.290	2.276	2.275	270	20.621	20.465	20.406	20.401
70	3.216	3.192	3.174	3.173	280	21.469	21.306	21.247	21.242
80	4.122	4.092	4.069	4.068	290	22.308	22.139	22.081	22.076
90	5.022	4.984	4.957	4.955	300	23.144	22.969	22.912	22.907
100	5.916	5.872	5.841	5.839	310	23.974	23.792	23.736	23.731
110	6.804	6.753	6.714	6.712	320	24.796	24.607	24.552	24.547
120	7.685	7.628	7.590	7.588	330	25.618	25.424	25.370	25.365
130	8.564	8.500	8.459	8.456	340	26.433	26.232	26.180	26.174
140	9.439	9.368	9.324	9.321	350	27.245	27.038	26.987	26.981
150	10.309	10.232	10.186	10.183	360	28.049	27.837	27.788	27.782
160	11.174	11.091	11.042	11.039	370	28.856	28.637	28.590	28.584
170	12.036	11.947	11.896	11.893	380	29.651	29.426	29.382	29.376
180	12.900	12.802	12.749	12.746	390	30.449	30.218	30.176	30.170
190	13.760	13.655	13.601	13.598	400	31.249	31.011	30.971	30.965
200	14.619	14.508	14.463	14.450	410	32.073	31.829	31.782	31.786
210	15.479	15.362	15.305	15.302	420	32.861	32.611	32.577	32.581
220	16.340	16.215	16.157	16.153					

TABLE XX. — THERMOMETER No. 6166.

Reading in Millimeters on Stem.	Temperature on Mercurial Thermometer, 0 and 100° fixed.	Temperature on Mercurial Thermometer, 0 and 40° fixed.	Temperature on Air Thermometer.	Temperature on Absolute Scale from 0° C.	Reading in Millimeters on Stem.	Temperature on Mercurial Thermometer, 0 and 100° fixed.	Temperature on Mercurial Thermometer, 0 and 40° fixed.	Temperature on Air Thermometer.	Temperature on Absolute Scale from 0° C.
20	−.036°	−.036°	−.034°	−.034°	230	16.478°	16.356°	16.298°	16.294°
30	+.770	+.764	+.759	+.759	240	17.259	17.132	17.074	17.070
40	1.574	1.562	1.553	1.553	250	18.042	17.908	17.849	17.845
50	2.368	2.350	2.336	2.335	260	18.825	18.686	18.627	18.622
60	3.156	3.133	3.115	3.114	270	19.609	19.464	19.405	19.400
70	3.941	3.911	3.889	3.888	280	20.392	20.241	20.182	20.177
80	4.726	4.691	4.665	4.664	290	21.176	21.019	20.960	20.955
90	5.509	5.468	5.438	5.436	300	21.735	21.793	21.735	21.730
100	6.293	6.246	6.212	6.210	310	22.511	22.569	22.511	22.506
110	7.076	7.024	6.988	6.986	320	23.292	23.349	23.292	23.287
120	7.862	7.804	7.765	7.763	330	24.075	24.131	24.075	24.070
130	8.649	8.585	8.544	8.542	340	24.855	24.910	24.855	24.850
140	9.437	9.367	9.323	9.321	350	25.634	25.687	25.634	25.628
150	10.228	10.151	10.105	10.102	360	26.415	26.466	26.412	26.406
160	11.017	10.935	10.887	10.884	370	27.441	27.245	27.195	27.189
170	11.805	11.717	11.667	11.664	380	28.240	28.030	27.982	27.976
180	12.589	12.496	12.444	12.441	390	29.030	28.814	28.768	28.762
190	13.370	13.271	13.217	13.214	400	29.819	29.597	29.550	29.544
200	14.148	14.043	13.988	13.984	410	30.608	30.381	30.339	30.333
210	14.923	14.812	14.756	14.752	420	31.396	31.162	31.123	31.117
220	15.699	15.583	15.526	15.522	430	32.189	31.950	31.914	31.908

In using these tables a correction is of course to be made should the zero point change.

TABLE XXI. — CORRECTION OF KEW STANDARD TO THE ABSOLUTE SCALE.

Temperature C.	Correction in Degrees C.
0°	0
10°	−.03
20°	−.05
30°	−.06
40°	−.07
50°	−.07
60°	−.06
70°	−.04
80°	−.02
90°	−.01
100°	0

Appendix to Thermometry.

The last of January, 1879, Mr. S. W. Holman, of the Massachusetts Institute of Technology, came to Baltimore to compare some thermometers with the air thermometer; and by his kindness I will give here the results of the comparison which we then made together.

As in this comparison some thermometers made by Fastré in 1851 were used, the results are of the greatest interest.

The tables are calculated with the newest value for the coefficient of expansion of glass. The calibration of all the thermometers, except the two by Casella, has been examined, and found good. The Casella thermometers had no reservoir at the top, and could not thus be readily calibrated after being made. The Geissler also had none, but I succeeded in separating a column.

The absence of a reservoir at the top should immediately condemn a standard, for there is no certainty in the work done with it.

TABLE XXII.—Seventh Series.

Air Thermometer.	Original Readings.					Reduced Readings.				
	6163.	7334 Baudin.	Kew Standard No. 104.	32374 Casella.	Geissler.	6163 Reduced to Air Thermometer.	7334 Baudin.	Kew Standard No.104.	32374 Casella.	Geissler.
°0	*58.83	—.11	32.68	+.20	+.69	°0	°0	°0	°0	°0
†.43	63.5	....	33.60	.71	....	.52	....	.52	.51	....
6.08	113.0	....	43.65	6.33	....	6.08	....	6.11	6.13	....
12.68	171.55	12.59	55.47	12.91	13.42	12.65	12.73	12.68	12 70	12.82
20.49	242.0	20.48	69.55	20.77	21.29	20.49	20.63	20.57	20.56	20.74
24.55	278.8	24.50	76.90	24.80	25.33	24.54	24 66	24.61	24.59	24.81
29.51	323.9	29.49	85.88	29.80	30.32	29.52	29.66	29.61	29.58	29.83
39.45	413.1	39.43	103.72	39.76	40.22	39.47	39.62	39.53	39.54	39.80
39.15	410.7	39.15	103.23	39.48	39.98	39.20	39.34	39.26	39.26	39.56
51.17	....	51.10	124.84	51.49	51.83	....	51.32	51.29	51.26	51.49
61.12	....	61.05	142.73	61.47	61.69	....	61.29	61.24	61.23	61.41
70.74	....	70.57	159.87	71.00	71.14	....	70 83	70.78	70.76	70.92
80.09	....	79.74	176.50	80.31	80.25	....	80.02	80.04	80.06	80.10
80.39	....	80.15	177.23	80.74	80.66	....	80.43	80.44	80.49	80.51
89.95	....	89.63	194.35	90.22	90.11	....	89.93	89.97	89.97	90.03
89.92	....	89.59	194.22	90.18	90.06	....	89.89	89.90	89.93	89.98
100.00	....	99.69	212.37	100.06	99.32	....	100.00	100.00	100.00	100.00

* The original readings in ice were 58.68 and 58 45, to which .15 was added to allow for the pressure of water in the comparator. This, of course, gives the same final result as if .15 were subtracted from each of the other temperatures. No correction was made to the others.

† Probably some error of reading.

TABLE XXIII. — Eighth Series.

Air Thermometer.	Original Readings.					6163 Reduced to Air Thermometer.	Reduced Readings.			
	6163	376 Fastré.	7316 Baudin.	368 Fastré.	3235 Casella.		376 Fastré.	7316 Baudin.	368 Fastré.	3235 Casella.
0°	*58.60	111.3	—.23	87.6	32.80	0°	0°	0°	0°	0°
8.67	90.7	130.0	....	106.25	39.35	3.61	3.64	....	3.64	3.65
11.55	161.6	170.9	11.40	147.2	53.70	11.56	11.60	11.64	11.62	11.63
20.72	243.7	217.9	20.59	194.2	70.15	20.70	20.75	20.84	20.80	20.79
32.19	347.4	276.9	32.09	253.2	90.80	32.17	32.24	82.34	32.28	82.29
39.36	411.85	318.85	39.26	290.1	103.68	39.36	39.43	39.52	39.48	39.45
50.71	....	372.0	50.57	248.2	123.65	....	50.75	50.84	50.80	50.57
60.10	....	420.0	59.92	396.45	140.80	....	60.10	60.19	60.21	60.12
73.82	....	490.6	73.59	466.85	165.68	....	73.84	73.87	73.93	73.97
86.50	....	555.25	86.16	531.22	188.20	....	86.48	86.51	86.56	86.56
....	....	550.2	85.21	525.95	186.42	....	86.45	85.50	85.45	85.51
100.00	....	624.93	99.70	600.58	212.45	....	100.00	100.00	100.00	100.00

From these tables we would draw the inference that No. 6163 represents the air thermometer with considerable accuracy. At the same time, both tables would give a smaller value of m than I have used, and not very far from the value found before by *direct* comparison, namely, .00015.

The difference from using $m = .00018$ would be a little over 0°.01 C. at the 20° point.

All the other thermometers stand *above* the air thermometer, between 0 and 100°, by amounts ranging between about 0°.05 and 0°.35 C., *none* standing below. Indeed, no table has ever been published showing any thermometer standing *below* the air thermometer between 0 and 100°. By inference from experiments above 100° on crystal glass by Regnault, thermometers of this glass should stand below, but it never seems to have been proved by direct experiment. The Fastré thermometers are probably made of this glass, and my Baudins certainly contain lead; and yet these stand above, though only to a small amount, in the case of the Fastré's.

The Geissler still seems to retain its pre-eminence as having the greatest error of the lot.

The Baudin thermometers agree well together, but are evidently made from another lot of glass from the No. 6167 used before. These last two depart less from the air thermometer. The explanation is plain, as Baudin had manufactured more than one thousand ther-

---

* See note on preceding page.

mometers between the two, and so had probably used up the first stock of glass. And even glass of the same lot differs, especially as Regnault has shown that the method of working it before the blow-pipe affects it very greatly.

It is very easy to test whether the calorimeter thermometers are of the same glass as any of the others, by testing whether they agree with No. 6163 throughout the whole range of 40°. The difference in the values of m for the two kinds of glass will then be about .003 of the difference between them at 20°, the 0 and 40° points agreeing. The only difficulty is in calibrating or reading the 100° thermometers accurately enough.

The Baudin thermometers were very well calibrated, and were graduated to $\frac{1}{10}$° C., and so were best adapted to this kind of work. Hence I have constructed the following tables, making the 0 and 40° points agree.

TABLE XXIV.— COMPARISON OF 6163 AND THE BAUDIN STANDARDS.

6163 Mercurial 0 and 40° fixed.	7334 *	Difference.	6163 Mercurial 0 and 40° fixed.	7316.*	Difference.
0	0	0	0	0	0
12.699	12.673	+.026	11.609	11.584	+.025
20.547	20.553	—.006	20.762	20.746	+.016
24.604	24.567	+.037	32.203	32.211	—.008
29.564	29.550	+.014	39.358	39.358	0
39.337	39.337	0			

Taking the average of the two, it would seem that No. 6163 stood about .015 higher than the mean of 7334 and 7316 at the 20° point, or 6163 has a higher value of m by .000045 than the others.

These differ about .17 from the air thermometer at 40°, which gives the value of m about .000104. Whence m for 6163 is .00015, as we have found before by direct comparison with the air thermometer.

I am inclined to think that the former value, .00018, is too large, and to take .00015, which is the value found by direct comparison, as the true value. As the change, however, only makes at most a difference of 0°.01 at any one point, and as I have already used the previous value in all calculations, I have not thought it worth while to go over all my work again, but will refer to the matter again in the final results, and then reduce the final results to this value.

---

* A correction of 0°.01 was made to the zero points of these thermometers on account of the pressure of the water.

III.—CALORIMETRY.

(a) Specific Heat of Water.

The first observers on the specific heat of water, such as De Luc, completed the experiment with a view of testing the thermometer; and it is curious to note that both De Luc and Flaugergues found the temperature of the mixture less than the mean of the two equal portions of which it was composed, and hence the specific heat of cold water *higher* than that of warm.

The experiments of Flaugergues were apparently the best, and he found as follows:[*]—

3 parts of water at 0° and 1 part at 80° R. gave 19°.86 R.
2 parts of "   "   2 parts "   " 39°.81 R.
1 part of   "   "   3 parts "   " 59°.87 R.

But it is not at all certain that any correction was made for the specific heat of the vessel, or whether the loss by evaporation or radiation was guarded against.

The first experiments of any accuracy on this subject seem to have been made by F. E. Newmann in 1831.[†] He finds that the specific heat of water at the boiling point is 1.0127 times that at about 28° C. (22° R.).

The next observer seems to have been Regnault,[‡] who, in 1840, found the mean specific heat between 100° C. and 16° C. to be 1.00709 and 1.00890 times that at about 14°.

But the principal experiments on the subject were published by Regnault in 1850,[§] and these have been accepted to the present time. It is unfortunate that these experiments were all made by mixing water above 100° with water at ordinary temperatures, *it being assumed that water at ordinary temperatures changed little, if any*. An interpolation formula was then found to represent the results; and *it was assumed* that the same formula held at ordinary temperature, or even as low as 0° C. It is true that Regnault experimented on the subject at points around 4° C. by determining the specific heat of lead in water at various temperatures; but the results were not of sufficient accuracy to warrant any conclusions except that the variation was not great.

---

[*] Gehler, Phys. Wörterbuch, i. 641.
[†] Pogg. Ann., xxiii. 40.   [‡] Ibid., li. 72.
[§] Pogg. Ann., lxxix. 241; also, Rel. d. Exp., i. 729.

Boscha has attempted to correct Regnault's results so as to reduce them to the air thermometer; but Regnault, in *Comptes Rendus*, has not accepted the correction, as the results were already reduced to the air thermometer.

Hirn (*Comptes Rendus*, lxx. 592, 831) has given the results of some experiments on the specific heat of water at low temperatures, which give the absurd result that the specific heat of water increases about six or seven per cent between zero and 13°! The method of experiment was to immerse the bulb of a water thermometer in the water of the calorimeter, until the water had contracted just so much, when it was withdrawn. The idea of thus giving equal quantities of heat to the water was excellent, but could not be carried into execution without a great amount of error. Indeed, experiments so full of error only confuse the physicist, and are worse than useless.

The experiments of Jamin and Amaury, by the heating of water by electricity, were better in principle, and, if carried out with care, would doubtless give good results. But no particular care seems to have been taken to determine the variation of the resistance of the wire with accuracy, and the measurement of the temperature is passed over as if it were a very simple, instead of an immensely difficult matter. Their results are thus to be rejected; and, indeed, Regnault does not accept them, but believes there is very little change between 5° and 25°.

In Poggendorff's *Annalen* for 1870 a paper by Pfaundler and Platter appeared, giving the results of experiments around 4° C., and deducing the remarkable result that water from 0 to 10° C. varied as much as twenty per cent. in specific heat, and in a very irregular manner, — first decreasing, then increasing, and again decreasing. But soon after another paper appeared, showing that the results of the previous experiments were entirely erroneous.

The new experiments, which extended up to 13° C., seemed to give an increase of specific heat up to about 6°, after which there was apparently a decrease. It is to be noted that Geissler's thermometers were used, which I have found to depart more than any other from the air thermometer.

But as the range of temperature is very small, the reduction to the air thermometer will not affect the results very much, though it will somewhat decrease the apparent change of specific heat.

In the *Journal de Physique* for November, 1878, there is a notice of some experiments of M. von Münchausen on the specific heat of water. The method was that of mixture in an open vessel, where evaporation might interfere very much with the experiment. No reference is made to the thermometer, but it seems not improbable that it was one from Geissler; in which case the error would be very great, as the range was large, and reached even up to 70° C. The error of the Geissler would be in the direction of making the specific heat increase more rapidly than it should. The formula he gives for the specific heat of water at the temperature t is

$$1 + .000302\ t.$$

Assuming that the thermometer was from Geissler, the formula, reduced to the air thermometer, would become approximately

$$1 - .00009\ t + .0000015\ t^2.$$

Had the thermometer been similar to that of Recknagel, it would have been $1 + .000045\ t + .000001\ t^2$.

It is to be noted that the first formula would actually give a decrease of specific heat at first, and then an increase.

As all these results vary so very much from each other, we can hardly say that we know anything about the specific heat of water between 0 and 100°, though Regnault's results above that temperature are probably very nearly correct.

It seems to me probable that my results with the mechanical equivalent apparatus give the variation of the specific heat of water with considerable accuracy; indeed, far surpassing any results which we can obtain by the method of mixture. It is a curious result of those experiments, that at low temperatures, or up to about 30° C., the specific heat of water is about constant *on the mercurial thermometer* made by Baudin, but *decreases to a minimum at about 30° when the reduction is made to the air thermometer or the absolute scale*, or, indeed, the Kew standard.

' As this curious and interesting result depends upon the accurate comparison of the mercurial with the air thermometer, I have spent the greater part of a year in the study of the comparison, but have not been able to find any error, and am now thoroughly convinced of the truth of this decrease of the specific heat.' But to make certain, I have instituted the following independent series of investigations on the specific heat of water, using, however, the same thermometers.

The apparatus is shown in Fig. 4. A copper vessel, A, about $20^{cm.}$

in diameter and 23$^{cm.}$ high, rests upon a tripod. In its interior is a three-way stopcock, communicating with the small interior vessel B, the vessel A, and the vulcanite spout C. By turning it, the vessel B could be filled with water, and its temperature measured by the thermometer D, after which it could be delivered through the spout into the calorimeter. As the vessel B, the stopcock, and most of the spout, were within the vessel A, and thus surrounded by water, and as the vulcanite tube was very thin, the water could be delivered into the

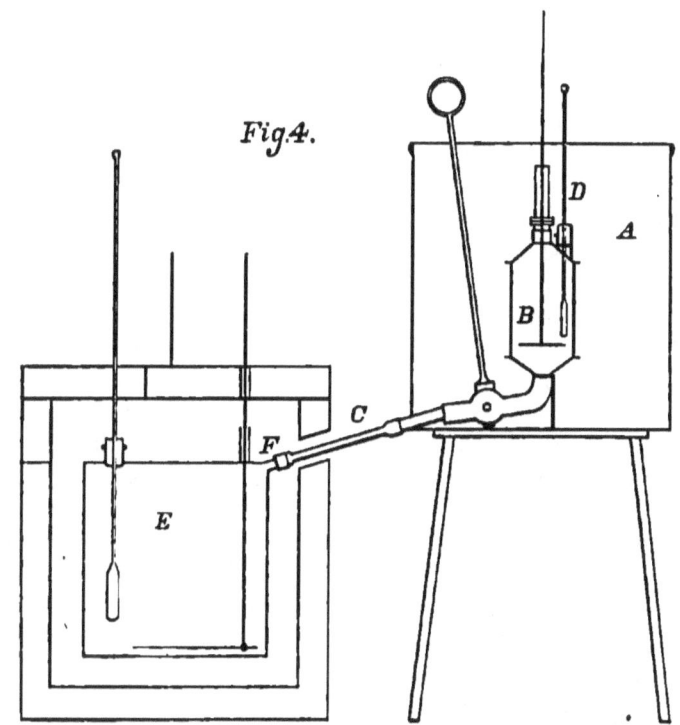

calorimeter without appreciable change of temperature. The proof of this will follow later.

The calorimeter, E, was of very thin copper, nickel-plated very thinly. A hole in the back at F allowed the delivery spout to enter, and two openings on top admitted the thermometers. A wire attached to a stirrer also passed through the top. The calorimeter had a capacity of about three litres, and weighed complete about 388.3 grammes. Its calorific capacity was estimated at 35.4 grammes. It rested on three vulcanite pieces, to prevent conduction to the jacket.

Around the calorimeter on all sides was a water-jacket, nickel-plated on its interior, to make the radiation perfectly definite.

The calorific capacity of the thermometers, including the immersed stem and the mercury of the bulb, was estimated as follows: $14^{cm.}$ of stem weighed about $3.8^{gr.}$, and had a capacity of $.8^{gr.}$; $10^{gr.}$ of mercury had a capacity of $.3^{gr.}$; total, $1.1^{gr.}$.

Often the vessel B was removed, and the water allowed to flow directly into the calorimeter.

The following is the process followed during one experiment at low temperatures. The vessel A was filled with clean broken ice, the opening into the stopcock being covered with fine gauze to prevent any small particles of ice from flowing out. The whole was then covered with cloth, to prevent melting. The vessel was then filled with water, and the two thermometers immersed to get the zero points. The calorimeter being about two thirds filled with water, and having been weighed, was then put in position, the holes corked up, and one thermometer placed in it, the other being in the melting ice. An observation of its temperature was then taken every minute, it being frequently stirred.

When enough observations had been obtained in this way, the cork was taken out of the aperture F and the spout inserted, and the water allowed to run for a given time, or until the calorimeter was full. It was then removed, the cork replaced, and the second thermometer removed from the ice to the calorimeter. Observations were then taken as before, and the vessel again weighed.

Two thermometers were used in the way specified, so that one might approach the final temperature from above and the other from below. But no regular difference was ever observed, and so some experiments were made with both thermometers in the calorimeter during the whole experiment.

The principal sources of error are as follows: —

1st. Thermometers lag behind their true reading. This was not noticed, and would probably be greater in thermometers with very fine stems like Geissler's. At any rate, it was almost eliminated in the experiment by using two thermometers.

2d. The water may be changed in temperature in passing through the spout. This was eliminated by allowing the water to run some time before it went into the calorimeter. The spout being very thin, and made of vulcanite, covered on the outside with cloth, it is not thought that there was any appreciable error. It will be discussed more at length below, and an experiment given to prove this.

3d. The top of the calorimeter not being in contact with the water, its temperature may be uncertain. To eliminate this, the calorimeter was often at the temperature of the air to commence with. Also the water was sometimes violently agitated just before taking the final reading, previous to letting in the cold water. Even if the temperature of this part was taken as that of the air, the error would scarcely ever be of sufficient importance to vitiate the conclusions.

4th. The specific heat of copper changes with the temperature. Unimportant.

5th. Some water might remain in the spout whose temperature might be different from the rest. This was guarded against.

6th. Evaporation. Impossible, as the calorimeter was closed.

7th. The introduction of cold water may cause dew to be deposited on the calorimeter. The experiments were rejected where this occurred.

The corrections for the protruding thermometer stem, for radiation, &c., were made as usual, the radiation being estimated by a series of observations before and after the experiment, as is usual in determining the specific heat of solids.

June 14, 1878. — First Experiment.

Time.	Ther. 6163.	Ther. 6166.	0 Points.	
41	296.75		6163, 57.9	Air, 21° C.
42	296.7		6165, 34.8	Jacket about 25° C.
43	296.7		6166, 20.5	
44	296.65			

44¼–44¾	Water running.		Calorimeter before	2043.0
46½	218.7	251.7	" after	2853.3
47¼	218.8	251.8	Water at 0° added	810.3
48½	218.9	252.0	Thermometer	1.1
			Total at 0°	811.4
Temperature before	296.6		Calorimeter before	2043.0
Correction for 0	+ .2		Weight of vessel	388.3
	296.8 = 26°.597		Water	1654.7
Correction for stem	+ .019		Capacity of calorim.	35.4
Initial temp. of calorimeter	26°.616		" thermom.	1.1
			Total capacity	1691.2

$218.6 + .2 = 218.8 = 17°.994$ $251.6 - 1 = 251.5 = 17°.962$
Correction for stem $-.006$ Correction for stem $-.006$
                                                           ———   ———
                                                      $17°.988$ $17°.956$

Mean temperature of mixture, $17°.972$.

$$\frac{\text{Mean specific heat } 0 - 18°}{\text{Mean specific heat } 18° - 27°} = \frac{1691.2 \times 8°.644}{811.4 \times 17°.972} = 1.0025.$$

June 14. — *Second Experiment.*

Calorimeter before 2016.3; temperature 361.4 by No. 6163.
    "   after 3047.0;   "   244.5 and 288.7.

Air, 21° C.; jacket about 27°.

$361.4 + .2 = 361.6 = 33°.803$, or $33°.863$ when corrected for stem.
$244.5 + .2 = 244.7 = 20°.865$; no correction for stem.
$288.7 - .1 = 288.6 = 20°.846$;   "   "

Mean, $20°.855$.

$$\frac{\text{Mean specific heat between 0 and } 21°}{\text{Mean specific heat between } 21° \text{ and } 34°} = 1.0062.$$

June 14. — *Third Experiment.*

Calorimeter before 1961.8; temperature 293.6 by No. 6166.
    "   after 3044.6;   "   243.7 and 213.0.

Air and jacket, about 18° C.

$393.6 - .1 = 393.5 = 29°.036$, or $29°.077$ when corrected for stem.
$243.7 - .1 = 243.6 = 17°.349$; no correction for stem.
$213.0 + .2 = 213.2 = 17°.374$;   "   "

Mean, $17°.361$.

$$\frac{\text{Mean specific heat between 0 and } 17°}{\text{Mean specific heat between } 17° \text{ and } 29°} = 1.0024.$$

It is to be observed that thermometer No. 6166 in all cases gave temperatures about 0°.02 or 0°.03 below No. 6163. This difference is undoubtedly in the determination of the zero points, as on June 15 the zero points were found to be 20.4 and 58.0. As one has gone up and the other down, the mean of the temperatures needs no correction.

June 15.

Calorimeter before 2068.2; temperature 364.6 by No. 6166.
" after 2929.2; " 249.7 and 217.7.

Air and jacket at about 22° C.

264.6 = 26°.766, or 26°.782 when corrected for stem.
249.7 = 17°.822, or 17°.812 " "
217.7 + .1 = 217.8 = 17°.884, or 17°.874 when corrected for stem.

Rejected on account of great difference in final temperatures by the two thermometers, which was probably due to some error in reading.

June 21.

Calorimeter before 2002.7; temperature 330.3 by No. 6163.
" after 3075.2; " 221.9 and 256.6.

Air and jacket, 21° C.

330.3 + .1 = 330.4 = 30°.321, or 30°.359 when corrected for stem.
221.9 + .1 = 222.0 = 18°.349, or 18°.343 " "
256.6 + .0 = 256.6 = 18°.358, or 18°.352 " "

Mean, 18°.347.

$$\frac{\text{Specific heat between 0 and 18°}}{\text{Specific heat between 18° and 30°}} = 1.0067.$$

June 21.

Calorimeter before 2073.8; temperature 347.8 by No. 6166.
" after 2986.8; " 234.5 and 206.6.

Air and jacket, about 21° C.

347.8 + .0 = 347.8 = 25°.457, or 25°.471 when corrected for stem.
234.5 + .0 = 234.5 = 16°.643, or 16°.636 " "
206.6 + .1 = 206.7 = 16°.651, or 16°.644 " "

Mean, 16°.640.

$$\frac{\text{Specific heat between 0 and 17°}}{\text{Specific heat between 17°. and 25°}} = .99971.$$

Rejected because dew was formed on the calorimeter.

A series was now tried with both thermometers in the calorimeter from the beginning.

June 25.

Calor. before 2220.3; temperat. 325.6 by No. 6166; 309.9 by No. 6165.
" after 3031.4; " 233.4 " " 224.6 " "
Air, 24°.2 C.; jacket, 23°.5.

325.6 + .0 = 325.6 = 23°.725, or 23°.726 when corrected for stem.
309.9 + .2 = 310.1 = 23°.739, or 23°.740 " "
233.4 + .0 = 233.4 = 16°.558, or 16°.545 " "
224.6 + .2 = 224.8 = 16°.562, or 16°.549 " "

Means, 23°.733 and 16°.547.

$$\frac{\text{Specific heat between 0 and 16°}}{\text{Specific heat between 16° and 24°}} = 1.0010.$$

June 25.

Calor. before 2278.6; temperat. 340.35 by No. 6166; 324.1 by No. 6165.
" after 3130.2; " 242.5 " " 232.8 " "
Air, 23°.5 C.; jacket, 22°.5.

340.35 + .0 = 340.35 = 24°.877, or 24°.881 when corrected for stem.
324.1 + .2 = 324.3 = 24°.899, or 24°.903 " "
242.5 + .0 = 242.5 = 17°.264, or 17°.253 " "
232.8 + .2 = 233.0 = 17°.261, or 17°.250 " "

$$\frac{\text{Specific heat between 0 and 17°}}{\text{Specific heat between 17° and 25°}} = 1.0027.$$

June 25.

Calor. before 2316.35; temperat. 386.1 by No. 6166; 368.4 by No. 6165.
" after 2966.90; " 295.4 " " 281.7 " "
Air, 23°.5 C.; jacket, 22°.5.

386.1 + .0 = 386.1 = 28°.455, or 28°.465 when corrected for stem.
368.4 + .2 = 368.6 = 28°.472, or 28°.482 " "
295.4 + .0 = 295.4 = 21°.374, or 21°.368 " "
281.7 + .2 = 281.9 = 21.°400, or 21°.394 " "

Means, 28°.473 and 21°.381.

$$\frac{\text{Specific heat between 0 and 21°}}{\text{Specific heat between 21° and 28°}} = 1.0045.$$

Two experiments were made on June 23 with warm water in vessel *A*, readings being taken of the temperature of the water, as it flowed out, by one thermometer, which was then transferred to the calorimeter as before.

June 23.

Water in A while running, 314.15 by No. 6163.

Calor. before 1530.9; temperat. 281.1 by No. 6166.
" after 2996.3; " 328.4 by No. 6166; 272.7 by No. 6163.

314.15 + .1 = 314.25 = 28°.526, or 28°.552 when corrected for stem.
281.1 + .0 = 281.1 = 20°.262, or 20°.258 " "
328.4 + .0 = 328.4 = 23°.945, or 23°.950 " "
272.7 + .1 = 272.8 = 23°.960, or 23°.966 " "

$$\frac{\text{Specific heat between } 20° \text{ and } 24°}{\text{Specific heat between } 24° \text{ and } 29°} = .9983.$$

June 23.

Water in A while running, 383.9 by No. 6163.

Calor. before 1624.9; temperat. 286.75 by 6166.
" after 3048.2; " 392.45 by 6166, and 318.1 by 6163.

383.9 + .1 = 384.0 = 36°.303, or 36°.357 when corrected for stem.
286.75 + .0 = 286.75 = 20°.702, or 20°.700 " "
392.45 + .0 = 392.45 = 28°.954, or 28°.980 " "
318.1 + .1 = 318.2 = 28°.964, or 28°.992 " "

$$\frac{\text{Specific heat between } 21° \text{ and } 29°}{\text{Specific heat between } 29° \text{ and } 36°} = .9954.$$

To test the apparatus, and also to check the estimated specific heat of the calorimeter, the water was almost entirely poured out of the calorimeter, and warm water placed in the vessel A, which was then allowed to flow into the calorimeter.

Water in A while running, 309.0 by No. 6163.

Calor. before 391.3; temperat. 314.5 by 6166.
" after 3129.0; " 308.3 by 6166, and 378.5 by 6163.

Air about 21° C.

Therefore, water lost 0°.078, and calorimeter gained 5°. Hence the capacity of the calorimeter is 39.

Another experiment, more carefully made, in which the range was greater, gave 35.

The close agreement of these with the estimated amount is, of course, only accidental, for they depend upon an estimation of only $0°.08$ and $0°.12$ respectively. But they at least show that the water is delivered into the calorimeter without much change of temperature.

A few experiments were made as follows between ordinary temperatures and $100°$, seeing that this has already been determined by Regnault.

Two thermometers were placed in the calorimeter, the temperature of which was about $5°$ below that of the atmosphere. The vessel B was then filled, and the water let into the calorimeter, by which the temperature was nearly brought to that of the atmosphere; the operation was then immediately repeated, by which the temperature rose about $5°$ above the atmosphere. The temperature of the boiling water was given by a thermometer whose $100°$ was taken several times.

As only the rise of temperature is needed, the zero points of the thermometers in the calorimeter are unnecessary, except to know that they are within $0°.02$ of correct.

June 18.

Temperature of boiling water, $99°.9$.

Calor. before 2684.7; temperat. 259.2 by 6166, and 248.3 by 6165.
" after 2993.2; " 381.0 " " 363.4 "

$259.3 = 18°.568$, or $18°.555$ when corrected for stem.
$248.3 = 18°.564$, or $18°.551$ " "
$381.0 = 28°.054$, or $28°.065$ " "
$363.4 = 28°.055$, or $28°.066$ " "

$$\frac{\text{Specific heat } 28° - 100°}{\text{Specific heat } 18° - 28°} = 1.0024.$$

Other experiments gave 1.0015 and 1.0060, the mean of all of which is 1.0033. Regnault's formula gives 1.005; but going directly to his experiments, we get about 1.004, the other quantity being for $110°$.

The agreement is very satisfactory, though one would expect my small apparatus to lose more of the heat of the boiling water than Regnault's. Indeed, for high temperatures my apparatus is much inferior to Regnault's, and so I have not attempted any further experiments at high temperatures.

My only object was to confirm by this method the results deduced from the experiments on the mechanical equivalent; and this I have done, for the experiments nearly all show that the specific heat of water *decreases* to about 30°, after which it increases. But the mechanical equivalent experiments give by far the most accurate solution of the problem; and, indeed, give it with an accuracy hitherto unattempted in experiments of this nature.

But whether water increases or decreases in specific heat from 0° to 30° depends upon the determination of the reduction to the air thermometer. *According to the mercurial thermometers Nos.* 6163, 6165, *and* 6166, *treating them only as mercurial thermometers, the specific heat of water up to* 30° *is nearly constant, but by the air thermometer, or by the Kew standard or Fastré, it decreases.*

Full and complete tables of comparison are published, and from them any one can satisfy himself of the facts in the case.

I am myself satisfied that I have obtained a very near approximation to absolute temperatures, and accept them as the standard. And by this standard the specific heat of water undoubtedly decreases from 0° to about 30°.

To show that I have not arrived at this result rashly, I may mention that I fought against a conclusion so much at variance with my preconceived notions, but was forced at last to accept it, after studying it for more than a year, and making frequent comparisons of thermometers, and examinations of all other sources of error.

However remarkable this fact may be, being the first instance of the decrease of the specific heat with rise of temperature, it is no more remarkable than the contraction of water to 4°. Indeed, in both cases the water hardly seems to have recovered from freezing. The specific heat of melting ice is infinite. Why is it necessary that the specific heat should instantly fall, and then recover as the temperature rises? Is it not more natural to suppose that it continues to fall even after the ice is melted, and then to rise again as the specific heat approaches infinity at the boiling point? And of all the bodies which we should select as probably exhibiting this property, water is certainly the first.

(b.) Heat Capacity of Calorimeter.

During the construction of the calorimeter, pieces of all the material were saved in order to obtain the specific heat. The calorimeter which Joule used was put together with screws, and with little or no solder. But in my calorimeter it was necessary to use solder, as it was of a much more complicated pattern. The total capacity of the solder used was only about $\frac{1}{800}$ of the total capacity including the water; and if we should neglect the whole, and call it copper, the error would be only about $\frac{1}{1000}$. Hence it was considered sufficient to weigh the solder before and after use, being careful to weigh the scraps. The error in the weight of solder could not possibly have been as great as ten per cent., which would affect the capacity only 1 part in 12,000.

To determine the nickel used in plating, the calorimeter was weighed before and after plating; but it weighed less after than before, owing to the polishing of the copper. But I estimated the amount from the thickness of a loose portion of the plating. I thus found the approximate weight of nickel, but as it was so small, I counted it as copper. The following are the constituents of the calorimeter: —

Thick sheet copper	25.1 per cent.
Thin " "	45.7 "
Cast brass	17.9 "
Rolled or drawn brass	5.7 "
Solder	4.0 "
Steel	1.6 "
	100.0 "
Nickel	.3 "

To determine the mean specific heat, the basket of a Regnault's apparatus was filled with the scraps in the above proportion, allowing the basket of brass gauze, which was very light, to count toward the drawn brass. The specific heat was then determined between 20° and 100°, and between about 10° and 40°. Between 20° and 100° the ordinary steam apparatus was used, but between 10° and 40° a special apparatus filled with water was used, the water being around the tube containing the basket, in the same manner as the steam is in the original apparatus. In the calorimeter a stirrer was used, so that the basket and water should rapidly attain the same temperature. The water was weighed before and after the experiment, to allow for evaporation. A correction of about 1 part in 1,000 was made, on account of the heat lost by the basket in passing from the apparatus to the calorimeter, in the 100° series, but no correction was made in the other series. The thermometers in the calorimeter were Nos. 6163 and 6166 in the different experiments.

The principal difficulty in the determination is in the correction for radiation, and for the heat which still remains in the basket after some time. After the basket has descended into the water, it commences to give out heat to the water; this, in turn, radiates heat; and the temperature we measure is dependent upon both these quantities.

Let T = temperature of the basket at the time t
" T' = " " " " " 0
" T'' = " " " " " ∞
" θ = " " water " t
" θ' = " " " " 0
" θ'' = " " " " ∞

$$\theta'' = T'''$$

We may then put approximately

$$T - T'' = (T' - T'')\varepsilon^{-\frac{t}{c}},$$

where c is a constant. But

$$\frac{T' - T''}{\theta'' - \theta'} = \frac{T' - T}{\theta - \theta'};$$

hence

$$\theta - \theta' \doteq (\theta'' - \theta')(1 - \varepsilon^{-\frac{t}{c}}).$$

To find c we have

$$c = \frac{1}{t}\log_\varepsilon \frac{\theta'' - \theta'}{\theta'' - \theta},$$

where θ'' can be estimated sufficiently accurately to find C' approximately.

These formulæ apply when there is no radiation. When radiation takes place, we may write, therefore, when t is not too small,

$$\theta - \theta' = (\theta'' - \theta')(1 - \varepsilon^{-\frac{t}{c}}) - C(t - t_0),$$

where C is a coefficient of radiation, and t_0 is a quantity which must be subtracted from t, as the temperature of the calorimeter does not rise instantaneously. To estimate t_0, T_a being the temperature of the air, we have, according to Newton's law of cooling,

$$C(t - t_0) = \frac{C}{\theta'' - T_a} \int_0^t (\theta - T_a)\,dt \text{ nearly,}$$

$$t_0 = c\frac{\theta'' - \theta'}{\theta'' - T_a} \text{ nearly,}$$

where it is to be noted that $\frac{C}{\theta'' - T_a}$ is nearly a constant for all values of $\theta'' - T_a$ according to Newton's law of cooling.

The temperature reaches a maximum nearly at the time

$$t_m = c \log_e \frac{\theta'' - \theta'}{c\,C};$$

and if θ_m is the maximum temperature, we have the value of θ'' as follows:

$$\theta'' = T'' = \theta_m + C(t_m + c - t_0);$$

$$T'' = \theta_m + C\left(t_m + c \frac{T_a - \theta'}{\theta'' - T_a}\right);$$

and this is the final temperature provided there was no loss of heat.

When the final temperature of the water is nearly equal to that of the air, C will be small, but the time t_m of reaching the maximum will be great. If a is a constant, we can put $C = a(\theta'' - T_a)$, and $C(t_m + c - t_0)$ will be a minimum, when

$$C = \frac{\theta'' - \theta'}{c}, \quad \text{or} \quad T_a = \theta'' - \frac{\theta'' - \theta'}{a\,c}.$$

That is, the temperature of the air must be lower than the temperature of the water, so that $T_a = \theta''$ as nearly as possible; but the formula shows that this method makes the corrections greater than if we make $T_a = \theta'$, the reason being that the maximum temperature is not reached until after an infinite time. It will in practice, however, be found best to make the temperature of the water at the beginning about that of the air. It is by far the best and easiest method to make all the corrections graphically, and I have constructed the following graphical method from the formulæ.

First make a series of measurements of the temperature of the water of the calorimeter, before and after the basket is dipped, together with the times. Then plot them on a piece of paper as in Fig. 5, making the scale sufficiently large to insure accuracy. Five or ten centimeters to a degree are sufficient.

$n\,a\,b\,c\,d$ is the plot of the temperature of the water of the calorimeter, the time being indicated by the horizontal line. Continue the line $d\,c$ until it meets the line $l\,a$. Draw a horizontal line through the point l. At any point, b, of the curve, draw a tangent and also a vertical line $b\,g$; the distance $e\,g$ will be nearly the value of the constant c in the formulæ. Lay off $l\,f$ equal to c, and draw the line $f\,h\,k$ through the point h, which indicates the temperature of the atmosphere or of the vessel surrounding the calorimeter. Draw a vertical line, $j\,k$, through the point k. From the point of maximum,

c, draw a line, jc, parallel to dm, and where it meets kj will be the required point, and will give the value of θ''. Hence, the rise of temperature, corrected for all errors, will be kj.

This method, of course, only applies to cases where the final temperature of the calorimeter is greater than that of the air; otherwise there will be no maximum.

In practice, the line dm is not straight, but becomes more and more nearly parallel to the base line. This is partly due to the constant decrease of the difference of temperature between the calorimeter and the air, but is too great for that to account for it. I have traced it to the thin metal jacket surrounding the calorimeter, and I must condemn, in the strongest possible manner, all such arrangements of calorimeters as have such a thin metal jacket around them. The jacket is

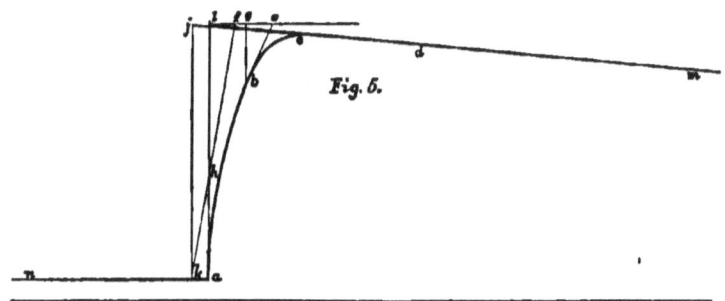

Fig. 5.

of an uncertain temperature, between that of the calorimeter and the air. When the calorimeter changes in temperature, the jacket follows it, but only after some time; hence, the heat lost in radiation is uncertain. The true method is to have a water jacket of constant temperature, and then the rate of decrease of temperature will be nearly constant for a long time.

The following results have been obtained by Mr. Jacques, Fellow of the University, though the first was obtained by myself. Corrections were, of course, made for the amount of thermometer stem in the air.

Temperature.	Mean Specific Heat.
24° to 100°	.0915
26° " 100°	.0915
25° " 100°	.0896
13° " 39°	.0895
14° " 38°	.0885
9° " 41°	.0910

To reduce these to the mean temperature of 0° to 40°, I have used the rate of increase found by Béde for copper. They then become, for the mean from 0° to 40°, —

.0897
.0897
.0878
.0893
.0883
.0906

Mean .0892 ± .00027

As the capacity of the calorimeter is about four per cent of that of the total capacity, including the water, this probable error is about $\frac{1}{1000}$ of the total capacity, and may thus be considered as satisfactory.

I have also computed the mean specific heat as follows, from other observers: —

Copper between 20° and 100° nearly.

.0949 Dulong.
.0935 }
.0952 } Regnault.
.0933 Béde.
.0930 Kopp.
.0940

This reduced to between 0° and 40° by Béde's formula gives .0922. Hence we have the following for the calorimeter: * —

	Per cent.	Specific Heat between 0° and 40° C.
Copper	91.4	.0922
Zinc	.7	.0896
Tin	3.6	.0550
Lead	2.7	.0310
Steel	1.6	.1110
		Mean .0895

The close agreement of this number with the experimental result can only be accidental, as the reduction to the air thermometer would decrease it somewhat, and so make it even lower than mine. How-

---

* The cast brass was composed of 28 parts of copper, 2 of tin, 1 of zinc, and 1 of lead. The rolled brass was assumed to have the same composition. The solder was assumed to be made of equal parts of tin and lead.

ever, the difference could not at most amount to more than 0.5 per cent, which is very satisfactory.

The total capacity of the calorimeter is reckoned as follows:—

<table>
<tr><td>Weight of calorimeter</td><td>3.8712 kilogrammes.</td></tr>
<tr><td>"   screws</td><td>.0016   "</td></tr>
<tr><td>"   part of suspending wires</td><td>.0052   "</td></tr>
<tr><td>Total weight</td><td>3.8780   "</td></tr>
</table>

Capacity $= 3.878 \times .0892 = .3459$ kilogrammes.

To this must be added the capacity of the thermometer bulb and several inches of the stem, and of a tube used as a safety valve, and we must subtract the capacity of a part of the shaft which was joined to the shaft turning the paddles. Hence,

$$\begin{aligned}&.3459\\&+.0011\\&+.0010\\&-.0010\\\hline\text{Capacity} = &.3470\end{aligned}$$

As this is only about four per cent of the total capacity, it is not necessary to consider the variation of this quantity with the temperature through the range from 0° to 40° which I have used.

IV. DETERMINATION OF EQUIVALENT.

(a.) Historical Remarks.

The history of the determination of the mechanical equivalent of heat is that of thermodynamics, and as such it is impossible to give it at length here.

I shall simply refer to the few experiments which *a priori* seem to possess the greatest value, and which have been made rather for the determination of the quantity than for the illustration of a method, and shall criticise them to the best of my ability, to find, if possible, the cause of the great discrepancies.

1. GENERAL REVIEW OF METHODS.

Whenever heat and mechanical energy are converted the one into the other, we are able by measuring the amounts of each to obtain the ratio. Every equation of thermodynamics proper is an equation between mechanical energy and heat, and so should be able to give us the mechanical equivalent. Besides this, we are able to measure a certain amount of electrical energy in both mechanical and heat units, and thus to also get the ratio. Chemical energy can be measured in heat units, and can also be made to produce an electric current of known mechanical energy. Indeed, we may sum up as follows the different kinds of energy whose conversion into one another may furnish us with the mechanical equivalent of heat. And the problem in general would be the ratio by which each kind of energy may be converted into each of the others, or into mechanical or absolute units.

    a. Mechanical energy.
    b. Heat.
    c. Electrical energy.
    d. Magnetic "
    e. Gravitation "
    f. Radiant "
    g. Chemical "
    h. Capillary "

Of these different kinds of energy, only the first five can be measured other than by their conversion into other forms of energy, although Sir William Thomson, by the introduction of such terms as "cubic mile of sunlight," has made some progress in the case of radiation. Hence for these five only can the ratio be known.

Mechanical energy is measured by the force multiplied by the distance through which the force acts, and also by the mass of a body multiplied by half the square of its velocity. Heat is usually referred to the quantity required to raise a certain amount of water so many degrees, though hitherto the temperature of the water and the reduction to the air thermometer have been almost neglected.

The energy of electricity at rest is the quantity multiplied by half the potential; or of a current, it is the strength of current multiplied by the electro-motive force, and by the time; or for all attractive forces varying inversely as the square of the distance, Sir William Thomson has given the expression

$$\frac{1}{8\pi} \int R^2 \, dv,$$

where R is the resultant force at any point in space, and the integral is taken throughout space.

These last three kinds of energy are already measured in absolute measure, and hence their ratios are accurately known. The only ratio, then, that remains is that of heat to one of the others, and this must be determined by experiment alone.

But although we cannot measure f, g, h in general, yet we can often measure off equal amounts of energy of these kinds. Thus, although we cannot predict what quantities of heat are produced when two atoms of different substances unite, yet, when the same quantities of the same substances unite to produce the same compound, we are safe in assuming that the same quantity of chemical energy comes into play.

According to these principles, I have divided the methods into direct and indirect.

Direct methods are those where b is converted directly or indirectly into a, c, d, or e, or *vice versa*.

Indirect methods are those where some kind of energy, as g, is converted into b, and also into a, c, d, or e.

In this classification I have made the arrangement with respect to the kinds of energy which are measured, and not to the intermediate steps. Thus Joule's method with the magneto-electric machine would be classed as mechanical energy into heat, although it is first converted into electrical energy. The table does not pretend to be complete, but gives, as it were, a bird's-eye view of the subject. It could be extended by including more complicated transformations; and, indeed, the symmetrical form in which it is placed suggests many other transformations. As it stands, however, it includes all methods so far used, besides many more.

In the table of indirect methods, the kind of energy mentioned first is to be eliminated from the result by measuring it both in terms of heat and one of the other kinds of energy, whose value is known in absolute or mechanical units.

It is to be noted that, although it is theoretically possible to measure magnetic energy in absolute units, yet it cannot be done practically with any great accuracy, and is thus useless in the determination of the equivalent. It could be thus left out from the direct methods without harm, as also out of the next to last term in the indirect methods.

TABLE XXV. — Synopsis of Methods for Obtaining the Mechanical Equivalent of Heat.

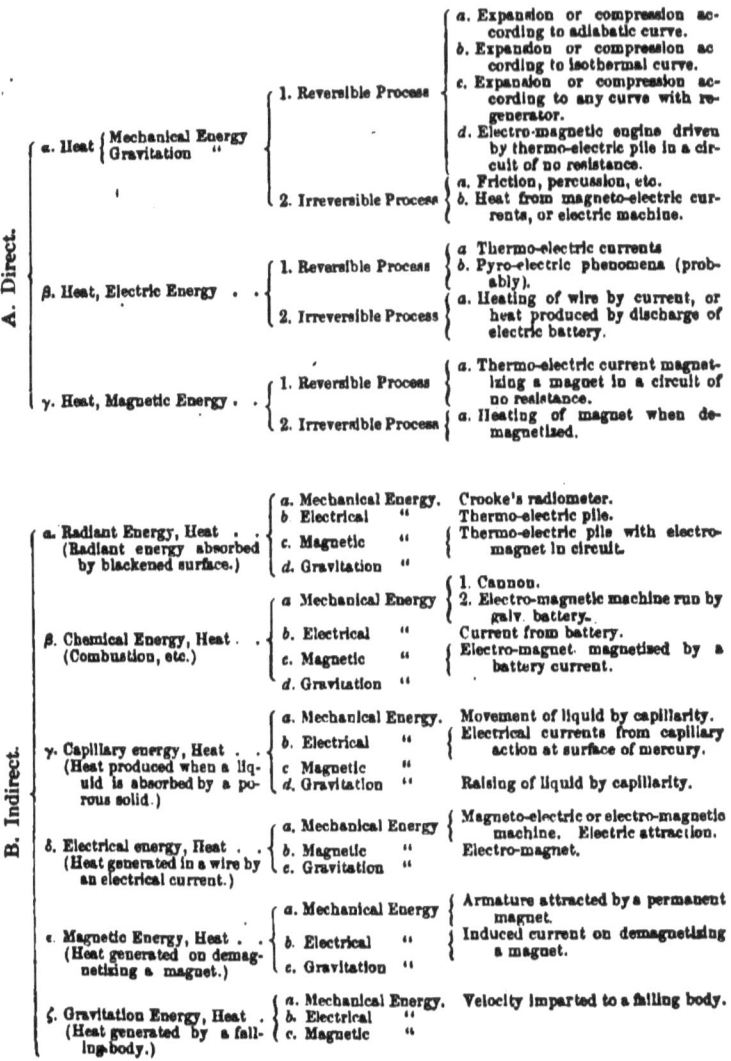

2. Results of Best Determinations.

On the basis of this table of methods I have arranged the following table, showing the principal results so far obtained.

In giving the indirect results, many persons have only measured one of the transformations required; and as it would lengthen out the

TABLE XXVI. — HISTORICAL TABLE OF EXPERIMENTAL RESULTS.

Method in General.				Method in Particular.	Observer.	Date.	Result.
A	a	1	a	Compression of air	Joule[2]	1845	443.8
				Expansion "	Joule[2]	1845	437.8
			b	Theory of gases (see below) . . .	. . . . .	. . . .	. . . .
			or	" vapors (see below) . .	. . . . .	. . . .	. . . .
			c	Experiments on steam-engine . .	Hirn[7]	1857	413.0
				" " " . .	Hirn[7]	1860-1	420-432
				Expansion and contraction of metals	Edlund[8]	1865	{ 443.6 430 1 428.3
A	a	2	a	Boring of cannon	Rumford[9]	1798	940ft.lbs.
				Friction of water in tubes. . . .	Joule[8]	1843	424.6
				" " in calorimeter .	Joule[4]	1845	488.3
				" " in calorimeter .	Joule[5]	1847	428.9
				" " in calorimeter .	Joule[6]	1850	423.9
				Friction of mercury in calorimeter	Joule[6]	1850	424.7
				" plates of iron	Joule[6]	1850	425.2
				" metals	Hirn[7]	1857	371.6
				" metals in mercury calor.	Favre[9]	1858	413.2
				" metals	Hirn[7]	1858	400-450
				Boring of metals	Hirn[7]	1858	425.0
				Water in *balance à frottement* . . .	Hirn[7]	1860-1	432.0
				Flow of liquids under strong pressure	Hirn[7]	1860-1	432.0
				Crushing of lead	Hirn[7]	1860-1	425.0
				Friction of metals	Puluj[12]	1876	426.6
				Water in calorimeter	Joule	1878	423.9
A	a	2	b	Heating by magneto electric currents	Joule[8]	1843	460.0
				Heat generated in a disc between the poles of a magnet . . .	Violle[10]	1870	{ 435.2 434.9 435.8 437.4
A	β	2	a	Heat developed in wire of known absolute resistance	Quintus Icilius,[11] also Weber	1857	399.7
				Do. do. do.	Lenz, also Weber	1859	{ 396.4 478.2
				Do. do. do.	Joule[13]	1867	429.5
				Do. do. do.	H.F.Weber[14]	1878	428.15
B	β	a	2	Diminishing of the heat produced in a battery circuit when the current produces work . . .	Joule[8]	1843	499.0
				Do. do. do.	Favre[15]	1858	443.0
B	β	b	1	Heat due to electrical current, electro-chemical equivalent of water = .009379, absolute resistance electro-motive force of Daniell cell, heat developed by action of zinc on sul. of copper	Weber, Boscha, Favre, and Silbermann	1857	432.1
				Heat developed in Daniell cell . .	Joule Boscha[12]	1859	419.5
				Electro-motive force of Daniell cell			

table very much to give the complete calculation of the equivalent from these selected two by two, I have sometimes given tables of these parts. As the labor of looking up and reducing these is very great, it is very possible that there have been some omissions.

I have taken the table published by the Physical Society of Berlin,[1] as the basis down to 1857, though many changes have been made even within this limit.

I shall now take up some of the principal methods, and discuss them somewhat in detail.

Method from Theory of Gases.

As the different constants used in this method have been obtained by many observers, I shall first give their results.

TABLE XXVII. — SPECIFIC HEAT OF GASES.

	Limit to Temperature	Approximate Temperature of Water	Temperature reduced to	Specific Heat.	
Air . . .	....	.....	Mercurial Thermometer	.2669	Delaroche and Bérard.
	20° to 210°	*14°.2	Air Thermometer	.23751[16]	Regnault.
	20° to 100°	20°	Mercurial Thermometer	.2389[17]	E. Wiedemann.
Hydrogen	....	.....	Mercurial Thermometer	3.2936	Delaroche and Bérard.
	15° to 200°	12°.2	Air Thermometer	3.4090[16]	Regnault.
	21° to 100°	21°	Mercurial Thermometer	3.410[17]	E. Wiedemann.

TABLE XXVIII. — COEFFICIENT OF EXPANSION OF AIR UNDER CONSTANT VOLUME.

	Taking Expansion of Mercury according to Regnault.	Taking Expansion of Mercury according to Wüllner's Re-calculation of Regnault's Experiments.
Regnault	.0036655	.0036687
Magnus	.0036678	.0036710
Jolly	.0036695	.0036727
Rowland	.0036675	.0036707
Mean	.0036676	.0036708

* Taking mean of results on page 101 of *Rel. des Exp.*, tom. ii.

TABLE XXIX.—Ratio of Specific Heats of Air.

Method.	Observer.	Date.	Ratio of Specific Heats.
Method of Clément & Désormes, globe 20 litres . .	Clément & Désormes[18]	1812 Published in 1819	1.354
Never fully published . . .	Gay-Lussac & Welter[19]	. . . .	1.3748
Method of Clément & Désormes	Delaroche & Bérard[41]	. . . .	1.249
Using Breguet thermometer .	Favre & Silbermann[28]	1853	1.421
Clément & Désormes, globe 39 litres	Masson[23]	1858	1.4196
Clément & Désormes . . .	Weisbach[21]	1859	1.4025
Clément & Désormes, globe 10 litres	Hirn[22]	1861	1.3845
Passage of gas from one vessel into another, globes 60 litres	Cazin[24]	1862	1.41
Pressure in globe changed by aspirator, globe 25 litres .	Dupré[25]	1863	. . . .
Heating of gas by electric current	Jamin & Richard[28]	1864	1.41
Clément & Désormes . . .	Tresca et Laboulaye[29]	1864	. . . .
Barometer under air-pump receiver of 6 litres . . .	Kohlrausch[26]	1869	1.302
Compression and expansion of gas by piston	Regnault	1871	Results lost in the siege of Paris.
Clément & Désormes with metallic manometer, globe 70 litres	Röntgen[27]	1873	1.4053
Compression of gas by piston .	Amagat[30]	1874	1.397

References. (Tables XXVI. to XXX.)

[1] Physical Society of Berlin, Fort. der Phys., 1858.

[2] Joule, Phil. Mag., ser. 3, vol. xxvi. See also Mec. Wärmeäquivalent, Gesammelte Abhandlungen von J. P. Joule, Braunschweig, 1872.

[3] Joule, Phil. Mag., ser. 3, vol. xxiii. See also 2 above.

[4] " " " " " xxvi. " "

[5] " " " " " xxvii. " "

[6] " " " " " xxxi. " "

[7] Hirn, Théorie Méc. de la Chaleur, ser. 1, 3me ed.

[8] Edlund, Pogg. Ann., cxiv. 1, 1865.

[9] Favre, Comptes Rend., Feb. 15, 1858; also Phil. Mag., xv. 406.

[10] Violle, Ann. de Chim., ser. 4, xxii. 64.

[11] Quintus Icilius, Pogg. Ann., ci. 69.

[12] Boscha, Pogg. Ann., cviii. 162.

[13] Joule, Report of the Committee on Electrical Standards of the B. A., London, 1873, p. 175.

[14] H. F. Weber, Phil. Mag., ser. 5, v. 30.

[15] Favre, Comptes Rend., xlvii. 599.

[16] Regnault, Rel. des Expériences, tom. ii.

[17] E Wiedemann, Pogg. Ann., clvii. 1.

TABLE XXX. — PRINCIPAL VALUES OF THE VELOCITY OF SOUND.

Number.	Observer	Date	Place.	Number of Observations.	Temperature Observed.	Velocity Observed.	Velocity reduced to 0° and Ordinary Air.	Velocity reduced to 0° and Dry Air.	** Velocity approximately reduced to 0° and Dry Air.
1	French Academy[31]	1738	France	.....	5° to 7°.5 C.	172.56 T.	332.9 m.	.....	332.6
2	Benzenberg[32]	1811	Dusseldorf	40	.....	.....	*333.7 m.	.....	332.7
3 {	Goldingham[33]	1821	India	120	83°.95 F.	1149.2 ft.	†333.0 cm.	.....	332.6
	"	1821	India	70	79°.9 F.	1131 5 ft.	†329.36 m.	.....	328.1
4	Bureau of Longitude[34]	1822	France	30	15°.9 C.	340.89 m.	331.36 m.	.....	330.8
5	Stampfer & Von Myrbach[35]	1822	Austria	88	9°.4 C.	.....	332.96 m.	.....	332.5
6 {	Moll & Van Beek[36]	1823	Holland	22 shots	11°.16 C.	340.37	333.62	$332.82	.....
	"	1823	Holland	14 shots	‡11°.00 C.	339.27	332.62	$381.91	.....
7	Parry & Foster[37]	1824–5	Port Bowen	61	−38°F. to +38°F.	.....	‖332.27	.....	332.0
8	Savart[38]	1839	.....	.....	5°.5 to 9°C.	336.5	¶332.2	.....	331.8
9	Bravais & Martins[39]	1844	Alps	34	8°.17 C.	338.01	383.11	332.37	.....
10	Regnault[40]	1864	France	149	2° to 20° C.	.....	.....	330.71	.....

* Reduced to 0° by empirical formula. † Wind calm. ‡ An error of 0°.21 C. was made in the original. See Shröder van der Kolk, Phil. Mag., 1865.
§ Moll & Van Beek found 332.049 m. at 0° in dry air. They used the coefficient .00375 to reduce. I take the numbers as recalculated by Shröder van der Kolk.
‖ Corrected for wind by Galbraith. ¶ Recalculated from Savart's results.
** I believe that I calculated these reduced numbers on the supposition that the air was rather more than half saturated with moisture.

Estimating the weight rather arbitrarily, I have combined them as follows: —

No.	Velocity at 0°. C. Dry Air.	Estimated Weight of Observation.
1	332.6	2
2	332.7	2
3	330.9	2
4	330.8	4
5	332.5	3
6	332.8	7
7	332.0	1
8	331.8	1
9	332.4	4
10	330.7	10
Mean	331.75	

Or, corrected for the normal carbonic acid in the atmosphere, it becomes 331.78 meters per second in dry pure air at 0° C.

---

[18] Clément et Désormes, Journal de Physique, lxxxix. 333, 1819.
[19] Laplace, Méc. Céleste, v. 125.
[20] Masson, Ann. de Chim. et de Phys., ser. 3, tom. liii.
[21] Weisbach, Der Civilingenieur, Neue Folge, Bd. v., 1859.
[22] Hirn, Théorie Méc. de la Chaleur, i. 111.
[23] Favre et Silbermann, Ann. de Chim., ser. 3, xxxvii. 1851.
[24] Cazin, Ann. de Chim., ser. 8, tom. lxvi.
[25] Dupré, Ann. de Chim., 3me ser., lxvii. 359, 1863.
[26] Kohlrausch, Pogg. Ann., cxxxvi. 618.
[27] Röntgen, Pogg. Ann., cxlviii. 603.
[28] Jamin and Richard, Comptes Rend., lxxi. 336.
[29] Tresca and Laboulaye, Comptes Rend., lviii. 358. Ann. du Conserv. des Arts et Métiers, vi. 365.
[30] Amagat, Comptes Rènd., lxxvii. 1325.
[31] Mém. de l'Acad. des Sci., 1738, p. 128.
[32] Benzenberg, Gilbert's Annalen, xlii. 1.
[33] Goldingham, Phil. Trans., 1823, p. 96.
[34] Ann. de Chim., 1822, xx. 210; also, Œuvres de Arago, Mém. Sci., ii. 1.
[35] Stampfer and Von Myrbach, Pogg. Ann., v. 496.
[36] Moll and Van Beek, Phil. Trans., 1824, p. 424. See also Shröder van der Kolk, Phil. Mag., 1865.
[37] Parry and Foster, Journal of the Third Voyage, 1824–5, Appendix, p. 86. Phil. Trans., 1828, p. 97.
[38] Savart, Ann. de Chim., ser. 2, lxxi. 20. Recalculated.
[39] Bravais and Martins, Ann. de Chim., ser. 3, xiii. 5.
[40] Regnault, Rel. des Exp., iii. 533.
[41] Delaroche and Bérard, Ann. de Chim., lxxxv. 72 and 113.
[42] Puluj, Pogg. Ann., clvii. 656.

From Regnault's experiments on the velocity in pipes I find by graphical means 331.4^m in free air, which is very similar to the above.

Calculation from Properties of Gases.

$K =$ specific heat of gas at constant pressure.
$k =$ "   "   "   "   volume.
$p =$ pressure in absolute units of a unit of mass.
$v =$ volume   "   "   "   "
$\mu =$ absolute temperature.
$J =$ Joule's equivalent in absolute measure.
$\gamma = \dfrac{K}{k}$.

General formula for all bodies: —

$$\gamma = \dfrac{1}{1 - \dfrac{\mu}{JK}\left(\dfrac{dp}{d\mu}\right)_v \left(\dfrac{dv}{d\mu}\right)_p},$$

$$\gamma = -\dfrac{V^2}{v^2}\left(\dfrac{dv}{dp}\right)_\mu;$$

$$\therefore J = \dfrac{\mu}{K}\left(\dfrac{dp}{d\mu}\right)_v \left(\dfrac{dv}{d\mu}\right)_p \dfrac{\gamma}{\gamma-1}.$$

Also,
$$J = -\dfrac{\mu}{K} \dfrac{\left(\dfrac{dv}{d\mu}\right)_p^2}{\left(\dfrac{dv}{dp}\right)_\mu + \dfrac{v^2}{V^2}}.$$

Application to gases; Rankine's formula is, —

$$pv = R\left(\mu - m\dfrac{\mu_0}{\mu}\dfrac{v_0}{v}\right),$$

$$\left(\dfrac{dp}{d\mu}\right)_v = \dfrac{p}{\mu}\left(1 + 2m\dfrac{\mu_0}{\mu}\dfrac{v_0}{v}\right),$$

$$\left(\dfrac{dv}{d\mu}\right)_p = \dfrac{v}{\mu}\left(1 + 3m\dfrac{\mu_0}{\mu^2}\dfrac{v_0}{v}\right),$$

$$\left(\dfrac{dv}{dp}\right)_\mu = -R\dfrac{\mu}{p^2} = -\dfrac{p_0 v_0}{\mu_0}\dfrac{\mu}{p^2}\left(1 + \dfrac{m}{\mu_0}\right).$$

If a_v is the coefficient of expansion between 0° and 100°, then

$$\mu_0 = \dfrac{1}{a_v}(1 + .00635\, m);$$

whence

$$J = \dfrac{pv\mu}{K} a'_p\, a'_v \left(\dfrac{\gamma}{\gamma-1}\right),$$

where a'_p and a'_v are the true coefficients of expansion at the given temperature;

$$\therefore J = \frac{pv}{K\mu}\left(1 + 5m\frac{\mu_0}{\mu^2}\frac{v_0}{v}\right)\frac{\gamma}{\gamma-1};$$

$$J = \frac{1}{K\mu}\frac{1 + 6m\frac{\mu_0}{\mu^2}\frac{v_0}{v}}{\frac{p_0 v_0 \mu}{p^2 v^2 \mu_0}\left(1+\frac{m}{\mu_0}\right) - \frac{1}{V^2}\left(\frac{v_0}{v}\right)^2}.$$

According to Thomson and Joule's experiments $m = 0°.33$ C. for air and about $2°.0$ for CO_2. Hence $\mu_0 = 272°.99$.

The equations should be applied to the observations directly at the given temperature, but it will generally be sufficient to use them after reduction to $0°$ C. Using $K = .2375$ according to Regnault for air, we have for the latitude of Baltimore, —

From Röntgen's value $\gamma = 1.4053$ $\dfrac{J}{g} = 430.3$.*

" Amagat's " 1.397 $\dfrac{J}{g} = 436.6$.

" velocity of sound 331.78^m per sec. $\dfrac{J}{g} = 429.6$.

Using Wiedemann's value for K, $.2389$, these become

$$\frac{J}{g} = 427.8; \qquad \frac{J}{g} = 434.0; \qquad \frac{J}{g} = 427.1.$$

As Wiedemann, however, used the mercurial thermometer, and as the reduction to the air thermometer would increase these figures from .2 to .8 per cent., it is evident that Regnault's value for K is the more nearly correct. I take the weights rather arbitrarily as follows: —

	Weight.	J.
Röntgen	3	430.3
Amagat	1	436.6
Velocity of sound	4	429.6
	Mean	430.7

And this is of course the value referred to water at $14°$ C. and in the latitude of Baltimore. My value at this point is 427.7.

---

* Röntgen gives the value 428.1 for the latitude of Paris as calculated by a formula of Shröder v. d. Kolk, and 427.3 from the formula for a perfect gas, and these both agree more nearly with my result than that calculated from my own formula.

This determination of the mechanical equivalent from the properties of air is at most very imperfect, as a very slight change in either γ or the velocity of sound will produce a great change in the mechanical equivalent.

From Theory of Vapors.

Another important method of calculating the mechanical equivalent of heat is from the equation for a body at its change of state, as for instance in vaporization. Let v be the volume of the vapor, and v' the volume of the liquid, and H the heat required to vaporize a unit of mass of the water; also let p be the pressure in absolute units, and μ the absolute temperature. Then

$$v - v' = \frac{JH}{\mu \left(\frac{dp}{d\mu}\right)_v}.$$

The quantity H and the relation of p to μ have been determined with considerable accuracy by Regnault. To determine J it is only required to measure the volume of saturated steam from a given weight of water; and the principal difficulty of the process lies in this determination, though the other quantities are also difficult of determination.

This volume can be calculated from the density of the vapor, but this is generally taken in the superheated state.

The experiments of Fairbairn and Tate [*] are probably the best direct experiments on the density of saturated vapor, but even those do not pretend to a greater accuracy than about 1 in 100. With Regnault's values of the other quantities, they give about Joule's value for the equivalent, namely 425. Hirn, Herwig, and others have also made the determination, but the results do not agree very well. Herwig even used a Giessler standard thermometer, which I have shown to depart very much from the air thermometer.

Indeed, the experiments on this subject are so uncertain, that physicists have about concluded to use this method rather for the determination of the volume of saturated vapors than for the mechanical equivalent of heat.

From the Steam-Engine and Expansion of Metals.

The experiments of Hirn on the steam-engine and of Edlund on the expansion and contraction of metals, are very excellent as illustrat-

---

[*] Phil. Mag., ser. 4, xxi. 230.

ing the theory of the subject, but cannot have any weight as accurate determinations of the equivalent.

From Friction Experiments.

Experiments of this nature, that is, irreversible processes for converting mechanical energy into heat, give by far the best methods for the determination of the equivalent.

Rumford's experiment of 1798 is only valuable from an historical point of view. Joule's results since 1843 undoubtedly give the best data we yet have for the determination of the equivalent. The mean of all his friction experiments of 1847 and 1850 which are given in the table is 425.8, though he prefers the smallest number, 423.9, of 1850. This last number is at present accepted throughout the civilized world, though there is at present a tendency to consider the number too small. But this value and his recent result of 1878 have undoubtedly as much weight as all other results put together.

As sources of error in these determinations I would suggest, first, the use of the mercurial instead of the air thermometer. Joule compared his thermometers with one made by Fastré. In the *Appendix to Thermometry* I give the comparison of two thermometers made by Fastré in 1850, with the air thermometer, as well as of a large number of others. From this it seems that all thermometers as far as measured stand *above* the air thermometer between 0° and 100°, and that the average for the Fastré at 40° is about 0°.1 C. Using the formula given in *Thermometry* this would produce an error of about 3 parts in 1,000 at 15° C., the temperature Joule used.

The specific heat of copper which Joule uses, namely, .09515, is undoubtedly too large. Using the value deduced from more recent experiments in calculating the capacity of my calorimeter, .0922, Joule's number would again be increased 13 parts in 10,000, so that we have, —

    Joule's value . . . .   423.9, water at 15°.7 C.
    Reduction to air thermometer . . $+$1.3
    Correction for specific heat of copper $+$ .5
    "   to latitude of Baltimore $+$ .5
                                         ─────
                                         426.2

It does not seem improbable that this should be still further increased, seeing that the reduction to the air thermometer is the smallest admissible, as most other thermometers which I have measured give greater correction, and some even more than three times as great as the one here used, and would thus bring the value even as high as 429.

One very serious defect in Joule's experiments is the small range of temperature used, this being only about half a degree Fahrenheit, or about six divisions on his thermometer. It would seem almost impossible to calibrate a thermometer so accurately that six divisions should be accurate to one per cent, and it would certainly need a very skilful observer to read to that degree of accuracy. Further, the same thermometer "A" was used throughout the whole experiment with water, and so the error of calibration was hardly eliminated, the temperature of the water being nearly the same. In the experiment on quicksilver another thermometer was used, and he then finds a higher result, 424.7, which, reduced as above, gives 427.0 at Baltimore.

The experiments on the friction of iron should be probably rejected on account of the large and uncertain correction for the energy given out in sound.

The recent experiments of 1878 give a value of 772.55, which reduced gives at Baltimore 426.2, the same as the other experiment.

The agreement of these reduced values with my value at the same temperature, namely 427.3, is certainly very remarkable, and shows what an accurate experimenter Joule must be to get with his simple apparatus results so near those from my elaborate apparatus, which almost grinds out accurate results without labor except in reduction. Indeed, the quantity is the same as I find at about 20° C.

The experiments of Hirn of 1860–61 seem to point to a value of the equivalent higher than that found by Joule, but the details of the experiment do not seem to have been published, and they certainly were not reduced to the air thermometer.

The method used by Violle in 1870 does not seem capable of accuracy, seeing that the heat lost by a disc in rapid rotation, and while carried to the calorimeter, must have been uncertain.

The experiments of Hirn are of much interest from the methods used, but can hardly have weight as accurate determinations. Some of the methods will be again referred to when I come to the description of apparatus.

Method by Heat generated by Electric Current.

The old experiments of Quintus Icilius or Lenz do not have any except historical value, seeing that Weber's measure of absolute resistance was certainly incorrect, and we now have no means of finding its error.

The theory of the process is as follows. The energy of electricity being the product of the potential by the quantity, the energy expended by forcing the quantity of electricity, Q, along a wire of resistance, R, in a second of time, must be $Q^2 R$, and as this must equal the mechanical equivalent of the heat generated, we must have $JH = Q^2 R t$, where H is the heat generated and t is the time the current Q flows.

The principal difficulty about the determination by this method seems to be that of finding R in absolute measure. A table of the values of the ohm as obtained by different observers, was published by me in my paper on the "Absolute Unit of Electrical Resistance," in the American Journal of Science, Vol. XV., and I here give it with some changes.

TABLE XXXI.

Date.	Observer.	Value of Ohm.	Remarks.
1849	Kirchoff	.88 to .90	Approximately.
1851	Weber	.95 to .97	Approximately.
1862	Weber	{ 1.088	From Thomson's unit.
		{ 1.075	From Weber's value of Siemens unit.
1863–4	B. A. Committee	{ 1.0000	Mean of all results.
		{ .993	Corrected by Rowland to zero velocity of coil.
1870	Kohlrausch	1.0193	
1873	Lorenz	.975	Approximately.
1876	Rowland	.9911*	From a preliminary comparison with the B. A. unit.
1878	H. F. Weber	1.0014	Using ratio of Siemens unit to ohm, .9536.

The ratio of the Siemens unit to the ohm is now generally taken at .9536, though previous to 1864 there seems to have been some doubt as to the value of the Siemens unit.

Since 1863–4, when units of resistance first began to be made with great accuracy, two determinations of the heat generated have been made. The first by Joule with the ohm, and the second by H. F. Weber, of Zurich, with the Siemens unit.

Each determination of resistance with each of these experiments gives one value of the mechanical equivalent. As Lorenz's result was only in illustration of a method, I have not included it among the exact determinations.

The result found by Joule was $J = 25187$ in absolute measure

---

- * Given .9012 by mistake in the other tables.

using feet and degrees F., which becomes 429.9 in degrees C. on a mercurial thermometer and in the latitude of Baltimore, compared with water at 18°.6 C.

TABLE XXXII. — EXPERIMENTS OF JOULE.

Observer.	Value of B. A. Unit.	Mechanical Equivalent from Joule's Exp.	Mechanical Equivalent reduced to Air Thermometer and corrected for Sp. Ht. of Copper.
B. A. Committee	1.0000	429.9	431.4
Ditto corrected by Rowland	.993	426.9	428.4
Kohlrausch	1.0193	438.2	439.7
Rowland	.9911	426.1	427.6
H. F. Weber	1.0014	430.5	432.0

The experiments of H. F. Weber[*] gave 428.15 in the latitude of Zurich and for 1° C. on the air thermometer and at a temperature of 18° C. This reduced to the latitude of Baltimore gives 428.45.

TABLE XXXIII.

Experiments of H. F. Weber.			Mean of Joule and Weber, giving Joule twice the Weight of Weber.
Observer.	Value of B. A. Unit.	Mechanical Equivalent of Heat from Weber's Experiments.	Mean Equivalent reduced to Air Thermometer in the Latitude of Baltimore.
B. A. Committee	1.000	427.9	430.2
Ditto corrected by Rowland	.993	424.9	427.2
Kohlrausch	1.0193	436.2	439.1
Rowland	.9911	424.1	426.4
H. F. Weber	1.0014	428.5	431.4

My own value at this temperature is 426.8, which agrees almost exactly with the fourth value from my own determination of the absolute unit.[†]

There can be no doubt that Joule's result is most exact, and hence I have given his results twice the weight of Weber's. Weber used a wire of about 14 ohms' resistance, and a small calorimeter holding only 250 grammes of water. This wire was apparently placed in the water without any insulating coating, and yet current enough was sent through

---

[*] Phil. Mag., 1878, 5th ser., v. 135.

[†] The value of the ohm found by reversing the calculation would be .992, almost exactly my value.

it to heat the water 15° during the experiment. No precaution seems to have been taken as to the current passing into the water, which Joule accurately investigated. Again, the water does not seem to have been continuously stirred, which Joule found necessary. And further, Newton's law of cooling does not apply to so great a range as 15°, though the error from this source was probably small. Furthermore, I know of no platinum which has an increase of coefficient of .001054 for 1° C., but it is usually given at about .003.

There can be no doubt that experiments depending on the heating of a wire give too small a value of the equivalent, seeing that the temperature of the wire during the heating must always be higher than that of the water surrounding it, and hence more heat will be generated than there should be. Hence the numbers should be slightly *increased*. Joule used wire of platinum-silver alloy, and Weber platinum wire, which may account for Weber's finding a smaller value than Joule, and Weber's value would be more in error than Joule's. Undoubtedly this is a serious source of error, and I am about to repeat an experiment of this kind in which it is entirely avoided. Considering this source of error, these experiments confirm both my value of the ohm and of the mechanical equivalent, and unquestionably show a large error in Kohlrausch's absolute value of the Siemens unit or ohm.

The experiments of Joule and Favre, where the heat generated by a current, both when it does mechanical work and when it does not, are very interesting, but can hardly have any weight in an estimation of the true value of the equivalent.

The method of calculating the equivalent from the chemical action in a battery, or the electro-motive force required to decompose any substance, such as water, is as follows.

Let E be such electro-motive force and c be the quantity of chemical substance formed in battery or decomposed in voltameter per second. Then total energy of current of energy per second is $E\,Q$, where Q is the current, or $c\,Q\,HJ$, where H is the heat generated by unit of c, or required to decompose unit of c. Hence, if the process is entirely reversible, we must have in either case

$$CHJ = E.$$

But the process is not always reversible, seeing that it requires more electro-motive force to decompose water than is given by a gas battery. This is probably due to the formation at first of some unstable compound like ozone. The process with a battery seems to be best, and we can thus apply it to the Daniell cell. The following quantities are mostly taken from Kohlrausch.

The quantity c has been found by various observers, and Kohlrausch* gives the mean value as .009421 for water according to his units (mg., mm., second system). Therefore for hydrogen it is .001047.

The quantity H can be observed directly by short-circuiting the battery, or can be found from experiments like those of Favre and Silbermann.

The electro-motive force E can be made to depend either upon the absolute measure of resistance, or can be determined, as Thomson has done, in electro-static units. In electro-magnetic units it is

	Siemens.	Ohms.	Absolute Measure according to my Determination.
After Waltenhofen	11.43	10.90	10.80×10^{10}
" Kohlrausch †	11.71	11.17	11.07×10^{10}

After Favre, 1 equivalent of zinc develops in the Daniell cell 23993 heat units;

$$\therefore \frac{J}{g} = \frac{E}{c H g}.$$

On the mg., mm., second system, we have $E = 10.935 \times 10^{10}$, $c = .001047$, $H = 23993$, $g = 9800.5$ at Baltimore.

$$\therefore \frac{J}{g} = 444160^{mm.} = 444.2 \text{ meters.}$$

Using Kohlrausch's value for absolute resistance, he finds 456.5, which is much more in error than that from my determination. I do not give the calculation from the Grove battery, because the Grove battery is not reversible, and action takes place in it even when no current flows.

Thomson finds the difference of potential between the poles of a Daniell cell in electro-static measure to be .00374 on the cm., grm., second system. ‡ Using the ratio 29,900 000 000$^{cm.}$ per second, as I have recently found, but not yet published, we have 111 800 000 on the electro-magnetic system or 11.18×10^{10} on the mm., mg., second system. This gives

$$\frac{J}{g} = 474.3 \text{ meters.}$$

---

* Pogg. Ann., cxlix. 179.
† Given by Kohlrausch, Pogg. Ann., cxlix. 182.
‡ Thomson, Papers on Electrostatics and Magnetism, p. 246.

General Criticism.

All the results so far obtained, except those of Joule, seem to be of the crudest description; and even when care was apparently taken in the experiment, the method seems to be defective, or the determination is made to rest upon the determination of some other constant whose value is not accurately known. Again, only one or two observers have compared their thermometers with the air thermometer, although I have shown in "Thermometry" that an error of more than one per cent may be made by this method. The range of temperatures is also small as a general rule and the specific heat of water is assumed constant.

Hence a new determination, avoiding these sources of error, seems to be imperatively demanded.

(b.) Description of Apparatus.

1. Preliminary Remarks.

As we have seen in the historical portion, the only experiments of a high degree of accuracy to the present time are those of Joule. Looked at from a general point of view, the principal defects of his method were the use of the mercurial instead of the air thermometer, and the small rate at which the temperature of his calorimeter rose.

In devising a new method a great rise of temperature in a short time was considered to be the great point, combined, of course, with an accurate measurement of the work done. For a great rise of temperature great work must be done, which necessitates the use of a steam-engine or other motive power. For the measurement of the work done, there is only one principle in use at present, which is, that the work transmitted by any shaft in a given time is equal to 2π times the product of the moment of the force by the number of revolutions of the shaft in that time.

In mechanics it is common to measure the amount of the force twisting the shaft by breaking it at the given point, and attaching the two ends together by some arrangement of springs whose stretching gives the moment. Morin's dynamometer is an example. Hirn[*] gives a method which he seems to consider new, but which is immediately recognized as Huyghens's arrangement for winding clocks with-

---

[*] Exposition de la Théorie Mécanique de la Chaleur, 8me éd., p. 18.

out stopping them. As cords and pulleys are used which may slip on each other, it cannot possess much accuracy. I have devised a method by cog-wheels which is more accurate, but which is better adapted for use in the machine-shop than for scientific experimentation.

But the most accurate method known to engineers for measuring the work of an engine is that of White's friction brake, and on this I have based my apparatus. Hirn was the first to use this principle in determining the mechanical equivalent of heat. In his experiment a horizontal axis was turned by a steam-engine. On the axis was a pulley with a flat surface, on which rested a piece of bronze which was to be heated by the friction. The moment of the force with which the friction tended to turn the piece of bronze was measured, together with the velocity of revolution. This experiment, which Hirn calls a *balance de frottement*, was first constructed by him to test the quality of oils used in the industrial arts. He experimented by passing a current of water through the apparatus and observing the temperature of the water before and after passing through. He thus obtained a rough approximation to Joule's equivalent.

He afterwards constructed an apparatus consisting of two cylinders about 30$^{cm.}$ in diameter and 100$^{cm.}$ long, turning one within the other, the annular space between which could be filled with water, or through which a stream of water could be made to flow whose temperature could be measured before and after. The work was measured by the same method as before.

But in neither of these methods does Hirn seem to have recognized the principle of the work transmitted by a shaft being equal to the moment of the force multiplied by the angle of rotation of the shaft. In designing his apparatus, he evidently had in view the reproduction in circular motion of the case of friction between two planes in linear motion.

Since I designed my apparatus, Puluj [*] has designed an instrument to be worked by hand, and based on the principle used by Hirn. He places the revolving axis vertical, and the friction part consists of two cones rubbing together. But no new principle is involved in his apparatus further than in that used by Hirn.[†]

---

[*] Pogg. Ann., clvii. 487.

[†] Joule's latest results were published after this was written, and I was not aware that he had made this improvement until lately. The result of his experiment, however, reached me soon after, and I have referred to it in the paper, but I did not see the complete paper until much later.

In my apparatus one of the new features has been the introduction of the Joule calorimeter in the place of the friction cylinders of Hirn or the cones of Puluj. At first sight the currents and whirlpools in such a calorimeter might be supposed to have some effect; but when the motion is steady, it is readily seen that the torsion of the calorimeter is equal to that of the shaft, and hence the principle must apply.

This change, together with the other new features in the experiments and apparatus, has at once made the method one of extreme accuracy, surpassing all others very many fold.

2. General Description.

The apparatus was situated in a small building, entirely separate from the other University buildings, and where it was free from disturbances.

Fig. 6 gives a general view of the apparatus. To a movable axis, $a\,b$, a calorimeter similar to Joule's is attached, and the whole is suspended by a torsion wire, c. The shaft of the calorimeter comes out from the bottom, and is attached to a shaft, $e\,f$, which receives a uniform motion from the engine by means of the bevel wheels g and h. To the axis, $a\,b$, an accurately turned wheel, $k\,l$, was attached, and the moment of the force tending to turn the calorimeter was measured by the weights o and p, attached to silk tapes passing around the circumference of this wheel in combination with the torsion of the suspending wire. To this axis was also attached a long arm, having two sliding weights, q and r, by which the moment of inertia could be varied or determined.

The number of revolutions was determined by a chronograph, which received motion by a screw on the shaft $e\,f$, and which made one revolution for 102 of the shaft. On this chronograph was recorded the transit of the mercury over the divisions of the thermometer.

Around the calorimeter a water jacket, $t\,u$, made in halves, was placed, so that the radiation could be estimated. A wooden box surrounded the whole, to shield the observer from the calorimeter.

The action of the apparatus is in general as follows. As the inner paddles revolve, the water strikes against the outer paddles, and so tends to turn the calorimeter. When this force is balanced by the weights $o\,p$, the whole will be in equilibrium, which is rendered stable by the torsion of the wire $c\,d$. Should any slight change take place in the velocity, the calorimeter will revolve in one direction or the other until the torsion brings it into equilibrium again. The amount of torsion read off on a scale on the edge of kl gives the correction to be added to or subtracted from the weights op.

One observer constantly reads the circle kl, and the other constantly records the transits of the mercury over the divisions of the thermometer.

A series extending over from one half to a whole hour, and recording a rise of 15° C. to perhaps 25° C., and in which a record was made for perhaps each tenth of a degree, would thus contain several hundred observations, from any two of which the equivalent of heat could be determined, though they would not all be independent. Such a series would evidently have immense weight; and, in fact, I estimate that, neglecting constant errors, a single series has more weight than all of Joule's experiments of 1849, on water, put together.*

The correction for radiation is inversely proportional to the ratio of the rate of work generated to the rate at which the heat is lost; and this for equal ranges of temperature is only $\frac{1}{50}$ as great in my measures as in Joule's; for Joule's rate of increase was about 0°.62 C. per hour, while mine is about 35° C. in the same time, and can be increased to over 45° C. per hour.

3. Details.

The Calorimeter.

Joule's calorimeter was made in a very simple manner, with few paddles, and without reference to the production of currents to mix up the water. Hence the paddles were made without solder, and were screwed together. Indeed, there was no solder about the apparatus.

But, for my purpose, the number of paddles must be multiplied, so that there shall be no jerk in the motion, and that the resistance may be great: they must be stronger, to resist the force from the engine, and they must be light, so as not to add an uncertain quantity to the calorific capacity. Besides this, the shape must be such as to cause the whole of the water to run in a constant stream past the thermometer, and to cause constant exchange between the water at the top and at the bottom.

---

* Forty experiments, with an average rise of temperature of 0°.56 F., equal to 0°.31 C., gives a total rise of 12°.4 C., which is only about two thirds the average of one of my experiments. As my work is measured with equal accuracy, and my radiation with greater, the statement seems to be correct.

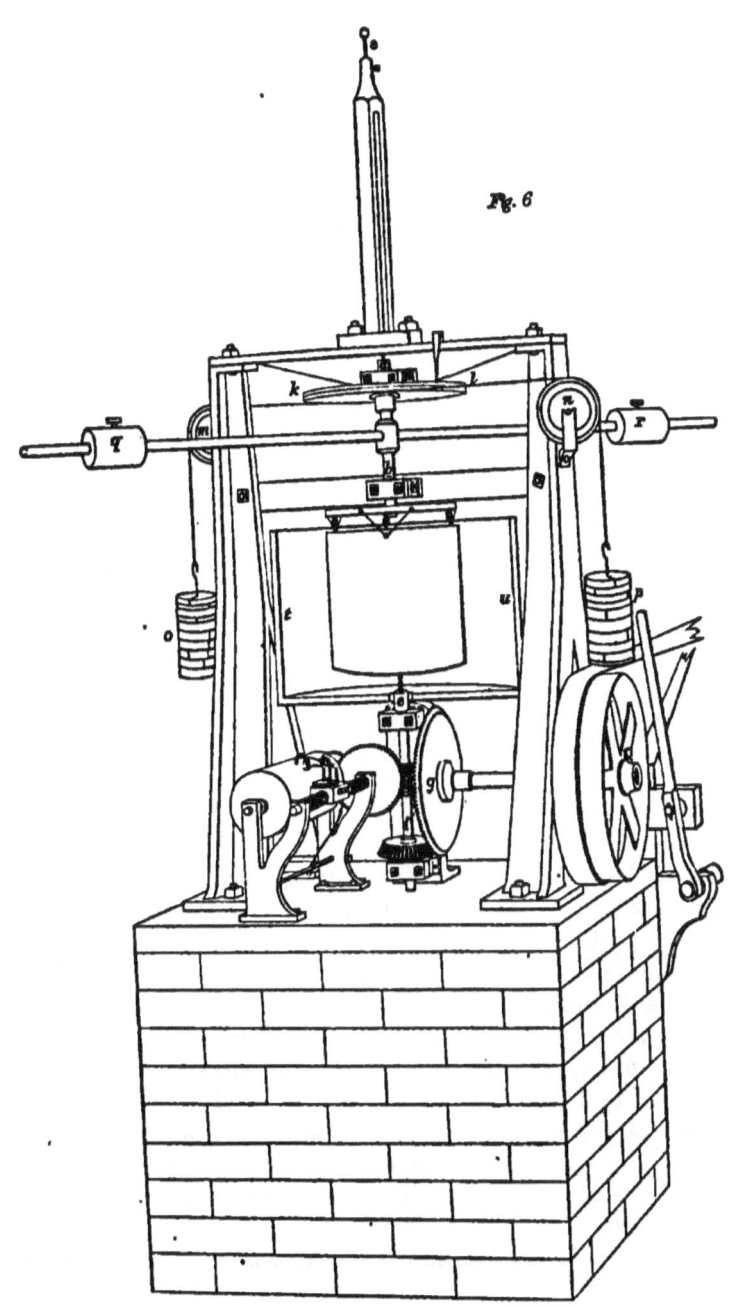

Fig. 6

Fig. 7 shows a section of the calorimeter, and Fig. 8 a perspective view of the revolving paddles removed from the apparatus, and with the exterior paddles removed from around it; which could not, however, be accomplished physically without destroying them.

To the axis $c\ b$, Fig. 7, which was of steel, and $6^{mm.}$ in diameter, a copper cylinder, $a\ d$, was attached, by means of four stout wires at e, and four more at f. To this cylinder four rings, g, h, i, j, were attached, which supported the paddles. Each one had eight paddles, but each ring was displaced through a small angle with reference to

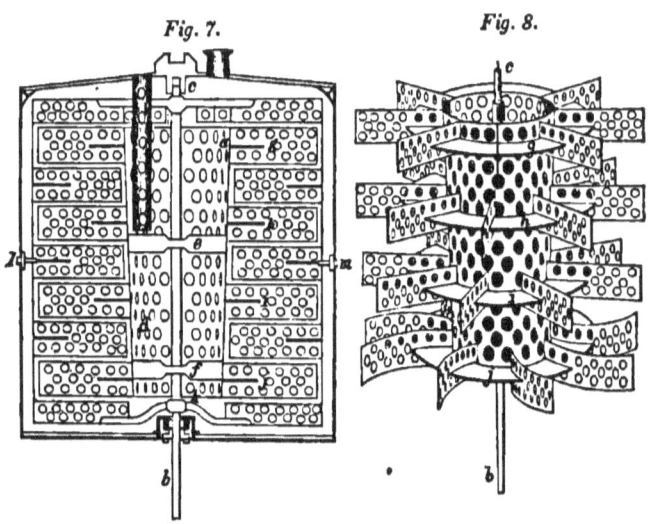

Fig. 7.   Fig. 8.

the one below it, so that no one paddle came over another. This was to make the resistance continuous, and not periodical. The lower row of paddles were turned backwards, so that they had a tendency to throw the water outwards and make the circulation, as I shall show afterwards.

Around these movable paddles were the stationary paddles, consisting of five rows of ten each. These were attached to the movable paddles by bearings, at the points c and k, of the shaft, and were removed with the latter when this was taken from the calorimeter. When the whole was placed in the calorimeter, these outer paddles were attached to it by means of four screws, l and m, so as to be immovable.

The cover of the calorimeter was attached to a brass ring, which was nicely ground to another brass ring on the calorimeter, and which could be made perfectly tight by means of a little white-lead paint. The shaft passed through a stuffing-box at the bottom, which was entirely within the outer surface of the calorimeter, so that the heat generated should all go to the water. The upper end of the shaft rested in a bearing in a piece of brass attached to the cover. In the cover there were two openings, — one for the thermometer, and the other for filling the calorimeter with water.

From the opening for the thermometer, a tube of copper, perforated with large holes, descended nearly to the centre of the calorimeter. The thermometer was in this sieve-like tube at only a short distance from the centre of the calorimeter, with the revolving paddles outside of it, and in the stream of water, which circulated as shown by the arrows.

This circulation of water took place as follows. The lower paddles threw the water violently outwards, while the upper paddles were

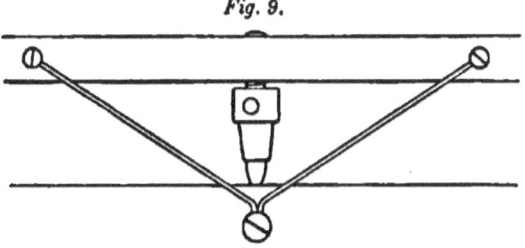

Fig. 9.

prevented from doing so by a cylinder surrounding the fixed paddles. The consequence was, that the water flowed up in the space between the outer shell and the fixed paddles, and down through the central tube of the revolving paddles. As there was always a little air at the top to allow for expansion, it would also aid in the same direction. These currents, which were very violent, could be observed through the openings.

The calorimeter was attached to a wheel, fixed to the shaft $a\,b$, by the method shown in Fig. 9. At the edge of the wheel, which was of the exact diameter of the calorimeter, two screws were attached, from which wires descended to a single screw in the edge of the calorimeter. Through the wheel, a screw armed with a vulcanite point pressed upon the calorimeter, and held it firmly. Three of these arrangements, at distances of 120°, were used. To centre the calorimeter, a piece of vulcanite at the centre was used. By this method of suspension very little heat could escape, and the amount could be allowed for by the radiation experiments.

The Torsion System.

The torsion wire was of such strength that one millimeter on the scale at the edge of the wheel signified 11.8 grammes, or about $\frac{1}{750}$ of the weights $o\ p$ generally used. There were stops on the wheel, so that it could not move through more than a small angle. The weights were suspended by very flexible silk tapes, 6^{mm} or 8^{mm} broad and 0.3^{mm} thick. They varied from 4.5^{k} to 8.5^{k} taken together. The shaft, $a\ b$, was of uniform size throughout, so that the wire c suspended the whole system, and no weight rested on the bearings.

The pulleys, m, n, Fig. 6, were very exactly turned and balanced, and the whole suspended system was so free as to vibrate for a considerable time. However, as will be shown hereafter, its freedom is of little consequence.

The Water Jacket.

Around the calorimeter, a water jacket, $t\ u$, was placed, so that the radiation should be perfectly definite. During the preliminary experiments a simple tin jacket was used, whose temperature was determined by two thermometers, one above and the other below, inserted in tubes attached to the jacket.

The Driving Gear.

The cog-wheels, g, h, were made by Messrs. Brown and Sharpe, of Providence, and were so well cut that the motion transmitted to the calorimeter must have been very uniform.

The Chronograph.

The cylinder of the chronograph was turned by a screw on the shaft $e f$, and received one revolution for 102 of the paddles; 155 revolutions of the cylinder, or 15,810 of the paddles, could be recorded, though, when necessary, the paper could be changed without stopping, and the experiment thus continued without interruption.

The Frame and Foundation.

The frame was very massive and strong, so as to prevent oscillation; and the whole instrument weighed about 500 pounds as nearly as could be estimated. It was placed on a solid brick pier, with a firm foundation in the ground. The trembling was barely perceptible to the hand when running the fastest.

The Engine.

The driving power was a petroleum engine, which was very efficient in driving the apparatus with uniformity.

The Balance.

For weighing the calorimeter, a balance capable of showing the presence of less than $\frac{1}{10}$ gramme with 15,000 grammes was used. The weights, however, by Schickert, of Dresden, were accurate among themselves to at least $5^{mg.}$ for the larger weights, and in proportion for the smaller. A more accurate balance would have been useless, as will be seen further on.

Adjustments.

There are few adjustments, and they were principally made in the construction.

In the first place, the shafts $a\,b$ and $e\,f$ must be on line. Secondly, the wheels $m\,n$ must be so adjusted that their planes are vertical, and that the tapes shall pass over them symmetrically, and that their edges shall be in the plane of the wheel $k\,l$.

Deviation from these adjustments only produced small error.

(c.) Theory of the Experiment.

1. ESTIMATION OF WORK DONE.

The calorimeter is constantly receiving heat from the friction, and is giving out heat by radiation and conduction. Now, at any given instant of time, the temperature of the whole of the calorimeter is not the same. Owing to the violent stirring, the water is undoubtedly at a very uniform temperature throughout. But the solid parts of the calorimeter cannot be so. The greatest difference of temperature is evidently soon after the commencement of the operation. But after some time the apparatus reaches a stationary state, in which, but for the radiation, the rise of temperature at all points would be the same. This steady state will be theoretically reached only after an infinite time; but as most of the metal is copper, and quite thin, and as the whole capacity of the metal work is only about four per cent of the total capacity, I have thought that one or two minutes was enough to allow, though, if others do not think this time sufficient, they can readily reject the first few observations of each series. When there is radiation, the stationary state will never be reached theoretically, though practically there is little difference from the case where there is no radiation.

The measurement of the work done can be computed as follows. Let M be the moment of the force tending to turn the calorimeter, and $d\theta$ the angle moved by the shaft. The work done in the time t will be $\int M\,d\theta$. If the moment of the force is constant, the integral is simply $M\theta$; but it is impossible to obtain an engine which runs with perfect steadiness, and although we may be able to calculate the integral, as far as long periods are concerned, by observation of the torsion circle, yet we are not thus able to allow for the irregularity during one revolution of the engine. Hence I have devised the following theory. I have found, by experiments with the instrument, that the moment of the force is very nearly, for high velocities at least, proportional to the square of the velocity. For rapid changes of the velocity, this is not exactly true, but as the paddles are very numerous in the calorimeter, it is probably very nearly true. We have then

$$M = C \left(\frac{d\theta}{dt}\right)^2,$$

where C is a constant. Hence the work done becomes

$$w = C \int \left(\frac{d\theta}{dt}\right)^2 d\theta = C \int \left(\frac{d\theta}{dt}\right)^3 dt.$$

As we allow for irregularities of long period by readings of the torsion circle, we can assume in this investigation that the mean velocity is constant, and equal to v_0. The form of the variation of the velocity must be assumed, and I shall put, without further discussion,

$$\frac{d\theta}{dt} = v_0 \left(1 + c \cos \frac{2\pi t}{a}\right).$$

We then find, on integrating from a to 0,

$$w = C v_0^3 a \left(1 + \tfrac{3}{2} c^2\right),$$

which is the work on the calorimeter during one revolution of the engine.

The equation of the motion of the calorimeter, supposing it to be nearly stationary, and neglecting the change of torsion of the suspending wire, is

$$\frac{m}{g} \frac{d^2\psi}{dt^2} - \frac{WD}{2} + C v_0^2 \left(1 + c \cos \frac{2\pi t}{a}\right)^2 = 0,$$

where m is the moment of inertia of the calorimeter and its attachments, ψ is the angular position of the calorimeter, W is the sum of the torsion weights, and D is the diameter of the torsion wheel. Hence,

$$\psi - \psi_0 = \frac{g}{m} \left\{ \tfrac{1}{2} t^2 [C v_0^2 (1 + \tfrac{1}{2} c^2) - W D] \right.$$
$$\left. + C v_0^2 \left[\frac{a^2 c^2}{16 \pi^2} \sin^2 \frac{2\pi t}{a} - \frac{a^2 c}{2 \pi^2} \left(\cos \frac{2 \pi t}{a} \right) \right] \right\}.$$

When $W D = 2 C v_0^2 (1 + \tfrac{1}{2} c^2)$, the calorimeter will merely oscillate around a given position, and will reach its maximum at the times $t = 0, \tfrac{1}{2} a, a,$ &c.

The total amplitude of each oscillation will be very nearly

$$\psi - \psi' = \frac{C v_0^2 g a^2 c}{\pi^2 m} = \frac{W D g a^2 c}{2 \pi^2 m}.$$

If x is the amplitude of each oscillation, as measured in millimeters, on the edge of the wheel of diameter D, we have $\psi - \psi' = \frac{2x}{D}$.

Hence, $$c = \tfrac{1}{2} \frac{m x n^2}{C D g},$$

where n is the number of revolutions of the engine per second.

Having found c in this way, the work will be, during any time,

$$w = \pi W D N (1 + c^2),$$

where N is the total number of revolutions of the *paddles*.

A variation of the velocity of ten per cent from the mean, or twenty per cent total, would thus only cause an error of one per cent in the equivalent.

Hence, although the engine was only single acting, yet it ran easily, had great excess of power, and was very constant as far as long periods were concerned. The engine ran very fast, making from 200 to 250 revolutions per minute. The fly-wheel weighed about 220 pounds, and had a radius of $1\tfrac{1}{4}$ feet. At four turns per second, this gives an energy of about 3400 foot pounds stored in the wheel. The calorimeter required about one-half horse-power to drive it; and, assuming the same for the engine friction, we have about 140 foot pounds of work required per revolution. Taking the most unfavorable case, where all the power is given to the engine at one point, the velocity changes during the revolution about four per cent, or c would nearly equal .02, causing an error of 1 part in 2500 nearly. By means of the shaking of the calorimeter, I have estimated c as follows, the value of m being changed by changing the weight on the inertia bar, or taking it off altogether. The estimate of the shaking was made by two persons independently.

m.	z observed.	c calculated.
2,200,000 grms. cm.²	.6 mm.	.016
3,100,000 "	.36 "	.013
11,800,000 "	.13 "	.017
	Mean,	$c = .015$

causing a correction of 1 part in 5000.

Another method of estimating the irregularity of running is to put on or take off weights until the calorimeter rests so firmly against the stops that the vibration ceases. Estimated in this way, I have found a little larger value of c, namely, about .017.

But as one cannot be too careful about such sources of error, I have experimented on the equivalent with different velocities and with very different ways of running the engine, by which c was greatly changed, and so have satisfied myself that the correction from this source is inappreciable in the present state of the science of heat.

Hence I shall simply put for the work

$$w = \pi N W D,$$

in gravitation measure at Baltimore. To reduce to absolute measure, we must multiply by the force of gravity given by the formula

$$g = 9.78009 + .0508 \sin^2 \phi,$$

which gives 9.8005 meters per second at Baltimore. If the calorimeter moved without friction, no work would be required to cause it to vibrate back and forth, as I have described; but when it moves *with* friction, some work is required. When I designed the apparatus, I thus had an idea that it would be best to make it as immovable as possible by adding to its moment of inertia by means of the inertia bar and weights. But on considering the subject further, I see that only the excess of energy represented by $c^2 \pi N W D$ can be used in this way. For, when the calorimeter is rendered nearly immovable by its great moment of inertia, the work done on it is, as we have seen, $\pi N W D (1 + c^2)$; but if it had no inertia, it is evident that the work would be only $\pi N W D$. If, therefore, the calorimeter is made partially stationary, either by its moment of inertia or by friction, the work will be somewhere between these two, and the work spent in friction will be only so much taken from the error. Hence in the latter experiments the inertia bar was taken off, and then the calorimeter constantly vibrated through about half a millimeter on the torsion scale.

Besides this quick vibration, the calorimeter is constantly moving to the extent of a few millimeters back and forth, according to the varying velocity of the engine. As frequent readings were taken, these changes were eliminated. In very rare cases the weights had to be changed during the experiment; but this was very seldom.

The vibration and irregular motion of the calorimeter back and forth served a very useful purpose, inasmuch as it caused the friction of the torsion apparatus to act first in one direction and then in the other, so that it was finally eliminated. The torsion apparatus moved very freely when the calorimeter was not in position, and would keep vibrating for some minutes by itself, but with the calorimeter there was necessarily some binding. But the vibration made it so free that it would return quickly to its exact position of equilibrium when drawn aside, and would also quickly show any small addition to the weights. This was tried in each experiment.

To measure the heat generated, we require to know the calorific capacity of the whole calorimeter, and the rise of temperature which would have taken place provided no heat had been lost by radiation. The capacity of the calorimeter alone I have discussed elsewhere, finding the total amount equal to $.347^k$ of water at ordinary temperatures. The total capacity of the calorimeter is then $A + .347$, where A is the weight of water. Hence Joule's equivalent in absolute measure is

$$J = \frac{102 \pi n W D}{(A + .347)(t - t')} g,$$

where n is the number of revolutions of the chronograph, it making one revolution to 102 of the paddles.

The corrections needed are as follows: —

1st. Correction for weighing in air. This must be made to W, the cast-iron weights, and to $A + .347$, the water and copper of the calorimeter. If λ is the density of the air under the given conditions, the correction is $-.835 \lambda$.

2d. For the weight of the tape by which the weights are hung. This is $\frac{.0006}{W}$.

3d. For the expansion of torsion wheel, D' being the diameter at 20° C. This is $.000018 (t'' - 20°)$. Hence,

$$J = 102 \pi g \frac{n W D'}{(A + .347)(t - t')} (1 + .000018 (t'' - 20) + \frac{.0006}{W} - .835 \lambda),$$

where $t - t'$ is the rise of the temperature corrected for radiation.

2. Radiation.

The correction for radiation varies, of course, with the difference of temperature between the calorimeter and jacket; but, owing to the rapid generation of heat, the correction is generally small in proportion. The temperature generated was generally about $0°.6$ per minute. The loss of temperature per minute by radiation was approximately $.0014\,\theta°$ per minute, where θ is the difference of the temperature. This is one per cent for $10°.7$, and four per cent for $14°.2$. Generally, the calorimeter was cooler than the jacket to start with, and so a rise of about $20°$ could be accomplished without a rate of correction at any point of more than four per cent, and an average correction of less than two per cent. An error of ten per cent is thus required in the estimation of the radiation to produce an average error of 1 in 500, or 1 in 250 at a single point. The coefficients never differ from the mean more than about two per cent. The observations on the equivalent, being at a great variety of temperatures, check each other as to any error in the radiation.

The losses of heat which I place under the head of radiation include conduction and convection as well. I divide the losses of heat into the following parts: 1st. Conduction down the shaft; 2d. Conduction by means of the suspending wires or vulcanite points to the wheel above; 3d. True radiation; 4th, Convection by the air. To get some idea of the relative amounts lost in this way, we can calculate the loss by conduction from the known coefficients of conduction, and we can get some idea of the relative loss from a polished surface from the experiments of Mr. Nichol. In this way I suppose the total coefficient of radiation to be made up approximately as follows: —

Conduction along shaft	.00011
" " suspending wires	.00006
True radiation	.00017
Convection	.00106
Total	.00140

The conduction through the vulcanite only amounts to $.0000002$.

From this it would seem that three fourths of the loss is due to radiation and convection combined.

The last two losses depend upon the difference of temperature between the calorimeter and the jacket, but the first two upon the difference between the calorimeter and frame of the machine and the wheel respectively. The frame was *always* of very nearly the same temperature as the water jacket, but the wheel was usually slightly above it. At first its temperature was noted by a thermometer, and the loss to it computed separately; but it was found to be unnecessary, and finally the whole was assumed to be a function of the temperature of the calorimeter and of the jacket only.

At first sight it might seem that there was a source of error in having a journal so near the bottom of the calorimeter, and joined to it by a shaft. But if we consider it a moment, we shall see that the error is inappreciable; for even if there was friction enough in the journal to heat it as fast as the calorimeter, it would decrease the radiation only seven per cent, or make an average error in the experiment of only 1 in 700. But, in fact, the journal was very perfectly made, and there was no strain on it to produce friction; besides which, it was connected to a large mass of cast-iron which was attached to the base. Hence, as a matter of fact, the journal was not appreciably warmer after running than before, although tested by a thermometer. The difference could not have been more than a degree or so at most.

The warming of the wheel by conduction and of the journal by friction would tend to neutralize each other, as the wheel would be warmer and the journal cooler during the radiation experiment than the friction experiment.

The usual method of obtaining the coefficient of radiation would be to stop the engine while the calorimeter was hot, and observe the cooling, stirring the water occasionally when the temperature was read. This method I used at first, reading the temperature at intervals of about a half to a whole hour. But on thinking the matter over, it became apparent that the coefficient found in this way would be too small, especially at small differences of temperature; for the layer next to the outside would be cooled lower than the mean temperature, and the heat could only get to the outside by conduction through the water or by convection currents.

Hence I arranged the engine so as to run the paddles very slowly, so as to stir the water constantly, taking account of the number of the revolutions and the torsion, so as to compute the work. As I had foreseen, the results in this case were higher than by the other method. At low temperatures the error of the first method was fifteen per cent; but at high, it did not amount to more than about three to five per cent, and probably at *very* high temperatures it would almost vanish.

I do not consider it necessary to give all the details of the radiation experiments, but will merely remark that, as the calorimeter was nickel-plated, and as seventy-five per cent of the so-called radiation is due to convection by the air, the coefficients of radiation were found to be very constant under similar conditions, even after long intervals of time.

The experiments were divided into two groups; one when the temperature of the jacket was about 5° C., and the other when it averaged about 20° C.

The results were then plotted, and the mean curve drawn through them, from which the following coefficients were obtained. These coefficients are the loss of temperature per minute, and per degree difference of temperature.

TABLE XXXV. — COEFFICIENTS OF RADIATION.

Difference between Jacket and Calorimeter.	Jacket 5°.	Jacket 20°.
−5	.00138	.00134
0	.00135	.00130
+5	.00137	.00132
10	.00142	.00138
15	.00148	.00144
20	.00154	.00150
25	.00158	.00154

As the quantity of water in the calorimeter sometimes varied slightly, the numbers should be modified to suit, they being true when the total capacity of the calorimeter was 8.75 kil. The total surface of the calorimeter was about 2350 sq. cm., and the unit of time *one minute*. To compare my results with those of McFarlane and of Nichol given in the Proc. R. S. and Proc. R. S. E., I will reduce my results so that they can be compared with the tables given by Professor Everett in his " Illustrations of the Centimeter-Gramme-Second System of Units," pp. 50, 51.

The reducing factor is .0621, and hence the last results for the jacket at 20° C. become : —

TABLE XXXVI.

Difference of Temperature.	Coefficient of Radiation on the C. G. S. System.	McFarlane's Value.	Ratio.
0	.000081	.000168	2.07
5	.000082	.000178	2.17
10	.000086	.000186	2.16
15	.000089	.000193	2.17
20	.000093	.000201	2.16
25	.000096	.000207	2.15

The variation which I find is almost exactly that given by McFarlane, as is shown by the constancy of the column of ratios. But my coefficients are less than half those of McFarlane. This may possibly be due to the fact that the walls of McFarlane's enclosure were blackened, and to his surface being of polished copper and mine of polished nickel: his surface may also have been better adapted by its form to the loss of heat by convection. The results of Nichol are also much lower than those of McFarlane.

The fact that the coefficients of radiation are less with increased temperature of jacket is just contrary to what Dulong and Petit found for radiation. But as I have shown that convection is the principal factor, I am at a loss to check my result with any other observer. Dulong and Petit make the loss from convection dependent only upon the difference of temperature, and approximately upon the square root of the pressure of the gas. Theoretically it would seem that the loss should be less as the mean temperature rises, seeing that the air becomes less dense and its viscosity increases. Should we substitute density for pressure in Dulong's law, we should have the loss by convection inversely as the square root of the mean absolute temperature, or approximately the absolute temperature of the jacket. This would give a decrease of one per cent in the radiation for about 6°, which is not far from what I have found.

To estimate the accuracy with which the radiation has been obtained is a very difficult matter, for the circumstances in the experiment are not the same as when the radiation was obtained. In the first place, although the water is stirred during the radiation, yet it is not stirred so violently as during the experiment. Further, the wheel above the calorimeter is warmer during radiation than during the experiment. Both these sources of error tend to give *too small* coefficients of radiation, and this is confirmed by looking over the final tables. But I have not felt at liberty to make any corrections based on the final results, as that would destroy the independence of the observations. But we are able thus to get the limits of the error produced.

During the preliminary experiments a water jacket was not used, but only a tin case, whose temperature was noted by a thermometer above and below. The radiation under these circumstances was larger, as the case was not entirely closed at the bottom, and so permitted more circulation of air.

8. Corrections to Thermometers, etc.

Among the other corrections to the temperature as read off from the thermometers, the correction for the stem at the temperature of

the air is the greatest. The ordinary formula for the correction is
.000156 n ($t - t''$). But, in applying this correction, it is difficult to
estimate n, the number of degrees of thermometer outside the calorimeter and at the temperature of the air, seeing that part of the stem
is heated by conduction. The uncertainty vanishes as the thermometer
becomes longer and longer, or rather as it is more and more sensitive.
But even then some of the uncertainty remains. I have sought to
avoid this uncertainty by placing a short tube filled with water about
the lower part of the thermometer as it comes out of the calorimeter.
The temperature of this was indicated by a thermometer, by aid of
which also the heat lost to the water by conduction through the thermometer stem could be computed; this, however, was very minute
compared with the whole heat generated, say 1 in 10,000.

The water being very nearly at the temperature of the air, the stem
above it could be assumed to be at the temperature of the air indicated
by a thermometer hung within an inch or two of it. The correction
for stem would thus have to be divided into two parts, and calculated
separately. Calculated in this way, I suppose the correction is perfectly certain to much less than one hundredth of a degree: the total
amount was seldom over one tenth of a degree.

Among the uncertain errors to which the measurement of temperature is subjected, I may mention the following: —

1. Pressure on bulb. A pressure of 60^{cm} of water produced a
change of about $0°.01$ in the thermometers. When the calorimeter
was entirely closed there was soon some pressure generated. Hence
the introduction of the safety-tube, — a tube of thin glass about
10^{cm} long, extending through a cork in the top of the calorimeter.
The top of the safety-tube was nearly closed by a cork to prevent
evaporation. Had the tube been shorter, water would have been
forced out, as well as air.

2. Conduction along stem from outside to thermometer bulb. To
avoid this, not only was the bulb immersed, but also quite a length of
stem. As this portion of the stem, as also the bulb, was surrounded
by water in violent motion, there could have been no large error from
this source. The immersed stem to the top of the bulb was generally
about 5^{cm} or more, and the stem only about $.8^{cm}$ in diameter.

3. The thermometer is never at the temperature of the water, because the latter is constantly rising; but we do not assume that it is
so in the experiment. We only assume that it lags behind the water
to the same amount at all parts of the experiment, and this is doubtless true.

To see if the amount was appreciable, I suddenly threw the apparatus out of gear, thus stopping it. The temperature was observed to continue rising about $0°.02$ C. Allowing $0°.01$ for the rise due to motion after the word "Stop" was given, we have about $0°.01$ C. as the amount the thermometer lagged behind the water.

4. Evaporation. A possible source of error exists in the cooling of the calorimeter by evaporation of water leaking out from it.

The water was always weighed before and after the experiment in a balance giving $\frac{1}{10}$ gramme with accuracy. The normal amount of loss from removal of thermometer, wet corks, &c. was about 1 gramme. The calorimeter was perfectly tight, and had no leakage at any point in its normal state. Once or twice the screws of the stuffing-box worked loose, but these experiments were rejected.

The evaporation of 1 gramme of water requires about 600 heat units, which is sufficient to depress the temperature of the calorimeter about $0°.07$ C. As the only point at which evaporation could take place was through a hole less than $1^{mm.}$ diameter in the safety-tube, I think it is reasonable to assume that the error from this source is inappreciable. But to be doubly certain, I observed the time which drops of water of known weight and area, placed on the warm calorimeter, took to dry. From these experiments it was evident that it would require a considerable area of wet surface to produce an appreciable effect. This wet surface never existed unless the calorimeter was wet by dew deposited on the cool surface. To guard against this error, the calorimeter was never cooled so low that dew formed; it was carefully rubbed with a towel, and placed in the apparatus half an hour to an hour before the experiment, exposed freely to the air. The surface being polished, the slightest deposit of dew was readily visible. The greatest care was taken to guard against this source of error, and I think the experiment is free from it.

(d.) **Results.**

1. Constant Data.

Joule's equivalent in gravitation measure is of the dimensions of length only, being the height which water would have to fall to be heated one degree. Or let water flow downward with uniform velocity through a capillary tube impervious to heat; assuming the viscosity constant, the rate of variation of height with temperature will be Joule's equivalent.

Hence, besides the force of gravity the only thing required in absolute measure is some length. The length that enters the equation is the diameter of the torsion wheel. This was determined under a microscope comparator by comparison with a standard meter belonging to Professor Rogers of Harvard Observatory, which had been compared at Washington with the Coast Survey standards, as well as by comparison with one of our own meter scales which had also been so compared. The result was .26908 meter at 20° C.

To this must be added the thickness of the silk tape suspending the weights. This thickness was carefully determined by a micrometer screw while the tape was stretched, the screw having a flat end. The result was $.00031^m$.

So that, finally, $D' = .26939$ meter at 20° C. Separating the constant from the variable parts, the formula now becomes

$$\frac{J}{g} = \frac{86.324}{A + .347}\left(1 + .000018\ (t'' - 20) + \frac{.0006}{W} - .835\ \lambda\right)\frac{Wn}{t - t'}.$$

$$g = 9.8005 \text{ at Baltimore.}$$

It is unnecessary to have the weights exact to standard, provided they are relatively correct, or to make double weighings, provided the same scale of the balance is always used. For both numerator and denominator of the fraction contain a weight.

2. Experimental Data and Tables of Results.

In exhibiting the results of the experiments, it is much more satisfactory to compute at once from the observations the work necessary to raise $1^{kil.}$ of the water from the first temperature observed to each succeeding temperature. By interpolation in such a table we can then reduce to even degrees. To compare the different results I have then added to each table such a quantity as to bring the result at 20° about equal to 10,000 kilogramme meters.

The process for each experiment may be described as follows. The calorimeter was first filled with distilled water a little cooler than the atmosphere, but not so cool as to cause a deposit of dew. It was then placed in the machine and adjusted to its position, though the outer half of the jacket was left off for some time, so that the calorimeter should become perfectly dry; to aid which the calorimeter was polished with a cloth. The thermometer and safety-tube were also inserted at this time.

After half an hour or so, the chronograph was adjusted, the outer half of the jacket put in place, the wooden screen fixed in position, and all was ready to start. The engine, which had been running quietly for some time, was now attached, and the experiment commenced. First the weights had to be adjusted so as to produce equilibrium as nearly as possible.

The observers then took their positions. One observer constantly recorded the transit of the mercury over the divisions of the thermometer, making other suitable marks, so that the divisions could be afterwards recognized. He also read the thermometers giving the temperatures of the air, the bottom of the calorimeter thermometer, and of the wheel just above the calorimeter; and sometimes another, giving that of the cast-iron frame of the instrument.

The other observer read the torsion wheel once every revolution of the chronograph cylinder, recording the time by his watch. He also recorded on the chronograph every five minutes by his watch, and likewise stirred the water in the jacket at intervals, and read its temperature.

The recording of the time was for the purpose of giving the connecting link between the readings of the torsion circle and of the thermometer. This, however, as the readings were quite constant, had only to be done roughly, say to half a minute of time, though the records of time on the chronograph were true to about a second.

The thermometers to read the temperature of the water in the jacket were graduated to $0°.2$ C., but were generally read to $0°.1$ C., and had been compared with the standards. There was no object in using more delicate thermometers.

After the experiment had continued long enough, the engine was stopped and a radiation experiment begun. The last operation was to weigh the calorimeter again, after removing the thermometer and safety-tube, and also the weights which had been used.

The chronograph sheet, having then been removed from the cylinder, had the time records identified and marked, as well as the thermometer records. Each line of the chronograph record was then numbered arbitrarily, and a table made indicating the stand of the thermometer and the number of the revolutions and fractions of a revolution as recorded on the chronograph sheet. The times at which these temperatures were reached was also found by interpolation, and recorded in another column.

From the column of times the readings of the torsion circle could be identified, and so all the necessary data would be at hand for calculating the work required to raise the temperature of one kilogramme of the water from the first recorded temperature to any succeeding temperature.

As these temperatures usually contained fractions, the amount of work necessary to raise one kilogramme of the water to the even degrees could then be found from this table by interpolation. Joule's equivalent at any point would then be merely the difference of any two succeeding numbers; or, better, one tenth the difference of two numbers situated 10° apart, or, in general, the difference of the numbers divided by the difference of the temperatures.

It would be a perfectly simple matter to make the record of the torsion circle entirely automatic, and I think I shall modify the apparatus in that manner in the future.

It would take too much space to give the details of each experiment; but, to show the process of calculation, I will give the experiment of Dec. 17, 1878 as a specimen. The chronograph sheet, of course, I cannot give. The computation is at first in gravitation measure, but afterwards reduced to absolute measure.

The calorimeter before the experiment weighed 12.2733 kil.
" " after " " " 12.2716 "
Mean 12.2720 "
Weight of calorimeter alone 3.8721 "
∴ Water alone weighed 8.3999 "
.3470 "
Total capacity 8.7469 "

The correction for weighing in air was .835 λ = .00106.
The total term containing the correction is therefore .99878.

log 86.324 = 1.9361316
log .99878 = 1.9994698
1.9356014
log 8.7469 = .9418542
log const. factor = .9937472 = log 9.85706.

Hence the work per kilogramme is 9.85706 ΣWn in gravitation measure, the term ΣWn being used to denote the sum of products similar to Wn as obtained by simultaneous readings of torsion circle and records on chronograph sheet.

Zero of torsion wheel, 79.3^{mm}.
Value of 1^{mm} on torsion wheel .0118^{kil}.

The following were the records of time on the chronograph sheet: —

Time observed.	Revolutions of Chronograph.	Time calculated.
15	8.74	15.2
20	25.32	20.1
25	42.10	25.0
30	59.05	30.0
35	76.00	35.0
40	93.03	40.0
45	109.97	45.0
50	126.92	50.0
55	144.14	55.0

The times were calculated by the formula

$$\text{Time} = .294 \times \text{Revolutions} + 12.66,$$

which assumes that the engine moves with uniform velocity. As the principal error in using an incorrect interpolation formula comes from the calculation of the radiation, and as this formula is correct within a few seconds for all the higher temperatures, we can use it in the calculation of the times.

The records of the transits of the mercury over the divisions of the thermometer were nearly always made for each division, but it is useless to calculate for each. I usually select the even centimeters, and take the mean of the records for several divisions on each side.

While the mercury was rising $1^{cm.}$ on No. 6163, there would be about seven revolutions of the chronograph, and consequently seven readings of the torsion circle, each one of which was the average for a little time as estimated by the eye.

I have obtained more than thirty series of results, but have thus far reduced only fourteen, five of which are preliminary, or were made with the simple jacket instead of the water jacket, the radiation to which was much greater, as there was a hole at the bottom which allowed more circulation of the air. The mean of the preliminary results agrees so closely with the mean of the final results, that I have in the end given them equal weight.

On March 24th, the same thermometer was used for a second experiment directly after the first, seeing that the chronograph failed to work in the first experiment until 8° was reached. The error from this cause was small, as the first experiment only reached to 26° C., and hence there could have been no change of zero, as this is very nearly the temperature at which the thermometer was generally kept.

Having thus calculated the work in conjunction with the temperature, I have next interpolated so as to obtain the work at the even degrees. The tables so formed I have combined in two ways: first, I have added to the column of work in each table an arbitrary number, such as to make the work at 20° *about* 10,000, and have then combined them as seen in Table LI.; and, secondly, I have subtracted each number from the one 10° farther down the table, and divided the numbers so found by 10, thus obtaining the mechanical equivalent of heat.

In these tables four thermometers have been used, and yet they were so accurate that little difference can be observed in the experiments which can be traced to an error of the thermometer, although the Kew standard has some local irregularities. The *greatest* difference between any column of Table LI. and the general mean is only 10 kilogramme-meters, or 0.023 degree, and this includes all errors of calibration of thermometers, radiation, &c. This seems to me to be a very remarkable result, and demonstrates the surpassing accuracy of the method. Indeed, the limit of accuracy in thermometry is the only limit which we can at present give to this method of experiment. Hence the large proportional time spent on that subject.

The accuracy of the radiation is demonstrated, to some extent, by the agreement of the results obtained even with different temperatures of the jacket. But on close observation it seems apparent that the coefficients of radiation should be further increased as there is a tendency of the end figures in each series to become too high. This is exactly what we should suppose, as we have seen that nearly all sources of error tend in the direction of making the radiation too small. For instance, an error came from not stirring the water during the radiation, and there must be a small residual error from not stirring so fast during radiation as during the experiment. Besides this, some parts around the calorimeter were warm during the radiation which were cool during the experiment. And both of these make the correction for radiation too small. However, the error from this source is small, and cannot possibly affect the general conclusions. In each column of Tables LI. and LII. a dash is placed at the temperature of the jacket, and for fifteen degrees below this point the error in the radiation must produce only an inappreciable error in the equivalent: taking the observations within this limit as the standards, and rejecting the others, we should still arrive at very nearly the same conclusions as if we accepted the whole.

Most of the experiments are made with a weight of about 7.3$^{kil.}$ as everything seemed to work best with this weight. But for the sake of a test I have run the weight up to 8.6 and down to 4.4$^{kil.}$ by which the rate of generation of the heat was changed nearly three times.

By this the correction for the radiation and the error due to the irregularity of the engine are changed, and yet scarcely an appreciable difference in the results can be observed.

The tables explain themselves very well, but some remarks may be in order. Tables XXXVII. to L. inclusive are the results of fourteen experiments selected from the total of about thirty, the others not having been worked up yet, though I propose to do so at my leisure.

Table LI. gives the collected results. At the top of each column the date of the experiment and number of the thermometer are given, together with the approximate torsion weight and the rate of rise of temperature per hour. The dash in each column gives approximately the temperature of the jacket, and hence of the air. There are four columns of mean values, but the last, produced from the combination of the table by parts, is the best.

Table LII. gives the mechanical equivalent of heat as deduced from intervals of 10° on Table LI. The selection of intervals of 10° tends to screen the variation of the specific heat of water from view, but a smaller interval gives too many local irregularities. In taking the mean I have given all the observations equal weight, but as the Kew standard was only graduated to $\frac{1}{4}$° F. it was impossible to calibrate it so accurately as to avoid irregularities of 0°.02 C. which would affect the quantities 1 in 500. Hence, in drawing a curve through the results, as given in the last column, I have almost neglected the Kew, and have otherwise sought to draw a regular curve without points of inflection. The figures in the last column I consider the best.

Table LIII. takes the mean values as found in Tables LI. and LII., and exhibits them with respect to the temperatures on the different thermometers, to the different parts of the earth, and also gives the reduction to the absolute scale. I am inclined to favor the absolute scale, using $m = .00015$, as given in the Appendix to Thermometry, rather than .00018, as used throughout the paper.

Table LIV. gives what I consider the final result of the experiment. It is based on the result $m = .00015$ for the thermometers, and is corrected for the irregularity of the engine by adding 1 in 4000.

The minor irregularities are also corrected so that the results signify a smooth curve, without irregularity or points of contrary flexure. But the curve for the work does not differ more than three kilogramme-meters from the actual experiment at any point, and generally coincides with it to about one kilogramme-meter. These differences signify 0°.007 C. and 0°.002 C., respectively. The mechanical equivalent is for single degrees rather than for ten degrees, as in the other tables.

TABLE XXXVII. — First Series. — *Preliminary.*
January 16, 1878. . Jacket and Air about 14° C.

Thermometer No. 6168.	Time.	Correction.		Corrected Temperature.	Revolutions of Chronograph Σ n.	Mean Weight W.	Work per Kilogramme.	Temperature.	Work per Kilogramme.	Work per Kilogramme + 5380
		Stem.	Rad.							
140	52.0	—.005	0	9.185	5.485	7.509	0	°	....	....
160	56.0	—.003	—.017	11.412	18.023	7.478	951	10	348	5728
180	59.2	0	—.022	13.650	30.652	7.442	1906	11	775	6155
203	63.4	+.006	—.015	16.230	45.329	7.394	3010	12	1202	6582
220	66.5	+.011	—.001	18.137	56.241	7.364	3825	13	1629	7009
240	70.2	+.020	+.027	20.392	69.158	7.354	4786	14	2056	7436
259	74.0	+.028	+.067	22.538	81.484	7.292	5702	15	2484	7864
289	80.0	+.045	+.161	25.943	101.214	....	7156	16	2912	8292
...	...	....	....	....	....	....	....	17	3340	8720
...	...	....	....	....	....	....	....	18	3767	9147
...	...	....	....	....	....	....	....	19	4193	9573
...	...	....	....	....	....	....	....	20	4610	9999
...	...	....	....	....	....	....	....	21	5048	10428
...	...	....	....	....	....	....	....	22	5472	10852
...	...	....	....	....	....	....	....	23	5899	11279
...	...	....	....	....	....	....	....	24	6326	11706
...	...	....	....	....	....	....	....	25	6753	12133
...	...	....	....	....	....	....	....	26	7180	12560

TABLE XXXVIII. — Second Series. — *Preliminary.*
March 7, 1878. Jacket 18°.5 to 22°.5. Air about 21° C.

Thermometer No. 6168.	Time.	Correction.		Corrected Temperature.	Revolutions of Chronograph Σ n.	Mean Weight W.	*Work per Kilogramme = Σ 10.060 W n.	Temperature.	Work per Kilogramme.	Work per Kilogramme + 6812.
		Stem.	Rad.							
170	19.9	—.016	0	12.537	5.03	7.737	0	13°	198	7010
180	...	....	....	13.646	11.12	7.710	474	14	625	7437
190	...	....	....	14.755	17.22	7.666	947	15	1052	7864
200	...	....	....	15.863	23.36	7.642	1421	16	1480	8292
210	26.8	—.010	—.036	16.972	29.55	7.641	1897	17	1909	8721
220	...	....	....	18.085	35.70	7.630	2369	18	2333	9145
230	...	....	....	19.196	41.90	7.611	2845	19	2761	9573
240	...	....	....	20.305	48.09	7.600	3319	20	3189	10001
250	33.8	+.003	—.036	21.419	54.30	7.596	3794	21	3615	10427
260	...	....	....	22.533	....	7.582	....	22	4041	10853
270	...	....	....	23.642	66.69	7.552	4740	23	4467	11279
280	...	....	....	24.754	72.92	7.547	5213	24	4892	11704
290	40.8	+.020	—.001	25.867	79.16	7.576	5687	25	5318	12130
300	...	....	....	26.990	85.42	....	6164	26	5744	12556

* In the calculation of this column, more exact data were used than given in the other two columns, seeing that the original calculation was made every 5 mm. of the thermometer. Hence the last figure may not always agree with the rest of the data.

OF ARTS AND SCIENCES. 181

TABLE XXXVIII. — *Continued.*

Thermometer No. 6163.	Time.	Correction.		Corrected Temperature.	Revolutions of Chronograph Σ n.	Mean Weight W.	Work per Kilogramme = Σ 10.060 W n.	Temperature.	Work per Kilogramme.	Work per Kilogramme + 6812.
		Stem.	Rad.							
310	...	....	....	28.119	91.67	7.611	6643	27	6168	12980
320	...	....	....	29.253	97.98	7.604	7125	28	6593	13405
330	47.8	+.044	+.078	30.393	104.28	7.611	7608	29	7017	13829
340	...	....	....	31.540	110.67	7.617	8097	30	7441	14253
350	51.4	....	....	32.689	117.12	7.602	8590	31	7867	14679
360	...	....	....	33.842	123.54	7.592	9081	32	8294	15106
370	55.0	+.072	+.184	84.998	130.04	7.576	9576	33	8722	15534
380	...	....	....	36.158	136.56	7.550	10071	34	9149	15961
390	58.7	+.588	+.261	37.321	143.08	7.550	10567	35	9577	16389
...	...	....	....	....	....	....	....	36	10004	16816
...	...	....	....	....	....	....	....	37	10430	17242

TABLE XXXIX. — Third Series. — *Preliminary.*
March 12, 1878. Jacket 13°.2 to 16°.6. Air about 15° C.

Thermometer No. 6163.	Time.	Correction.		Corrected Temperature.	Revolutions of Chronograph Σ n.	*Mean Weight W.	Work per Kilogramme = Σ 9.9690 W n.	Temperature.	Work per Kilogramme.	Work per Kilogramme + 7509.
		Stem.	Rad.							
205	28.0	0	0	14.368	3.156	⎫	0	. .	....	....
210	28.6	0	+.002	14.754	5.334	⎬ 7.5167	164	15	269	7868
220	29.9	....	....	15.529	9.770		495	16	696	8295
230	31.1	+.003	+.010	16.307	14.184	⎭	827	17	1122	8721
240	32.4	....	....	17.090	18.642	⎫	1160	18	1548	9147
250	33.6	+.009	+.021	17.875	23.080	⎬ 7.5462	1495	19	1975	9574
260	34.9	....	....	18.662	27.550	⎭	1831	20	2401	10000
270	36.2	+.014	+.038	19.452	32.014	⎫	2167	21	2828	10427
280	37.4	....	....	20.242	36.474	⎬ 7.5668	2504	22	3253	10852
290	38.7	+.019	+.055	21.029	40.924	⎭	2840	23	3676	11275
300	39.9	....	....	21.825	45.424	⎫	3179	24	4101	11700
310	41.2	+.024	+.089	22.619	49.838	⎬ 7.5875	3514	25	4526	12125
320	42.5	....	....	23.418	54.302	⎭	3853	26	4951	12550
330	43.7	+.030	+.120	24.220	58.844	⎫	4194	27	5378	12977
340	45.0	....	....	25.023	63.366	⎬ 7.5763	4536	28	5803	13402
350	46.3	+.038	+.159	28.825	67.874	⎭	4876	29	6226	13825
360	47.6	....	....	26.628	72.403	⎫	5219	30	6653	14252
370	48.9	+.047	+.202	27.438	76.987	⎬ 7.5872	5565	31	7078	14677
380	50.1	....	....	28.253	81.550	⎭	5910	..	....	....
390	51.4	+.056	+.251	29.069	86.100	⎫	6255	..	....	....
400	52.7	....	....	29.884	90.720	⎬ 7.5801	6604	..	....	....
410	54.0	+.066	+.304	30.703	95.316	⎭	6951	..	....	....
420	55.3	....	....	31.519	99.920	....	7299	..	....	....

* As this table was originally calculated for every 5 mm. on the thermometer, I have given the weights which were used to check the more exact calculation.

TABLE XL. — FOURTH SERIES. — *Preliminary.**

March 24, 1878. Jacket 5°.4 to 8°.2. Air about 6° C.

Thermometer No. 6168.	Time.	Correction.		Corrected Temperature.	Revolutions of Chronograph 1 n.	Mean Weight W.	Work per Kilogramme = 2 9.9816 W n.	Temperature.	Work per Kilogramme.	Work per Kilogramme + 4902.
		Stem.	Rad.							
130	27.4	+.002	0	8.071	42.364	7.471	0	8	—30	4872
140	29.2	....	....	9.204	48.898	7.446	485	9	398	5300
150	31.0	+.010	+.019	10.340	55.438	7.442	968	10	823	5725
160	32.9	....	....	11.480	62.066	7.405	1458	11	1252	6154
170	34.7	+.017	+.050	12.620	68.669	7.300	1944	12	1680	6582
180	36.6	....	....	13.763	75.330	7.398	2433	13	2107	7009
190	38.4	+.025	+.093	14.908	81.973	7.431	2921	14	2534	7436
200	40.3	....	....	16.054	88.507	7.429	3410	15	3960	8862
210	42.2	+.034	+.150	17.202	95.264	7.437	3902	16	3387	8289
220	44.2	....	....	18.350	101.941	7.433	4395	17	3815	8717
230	46.1	+.046	+.222	19.504	108.588		4886	18	4245	9147
240	...	....	....	....	....		....	19	4672	9574
250	...	....	....	....	....	} 7.4617	....	20	5098	10000
260	...	....	....	....	....		....	21	5524	10426
270	53.6	+.073	+.399	24.124	135.158	7.509	6855	22	5950	10852
280	55.7	....	....	25.288	141.803	7.502	7350	23	6376	11278
290	57.7	+.084	+.524	26.456	148.427	....	7844	24	6802	11704
...	..	....	....	....	....	....	....	25	7228	12130
...	..	....	....	....	....	....	....	26	7651	12553

TABLE XLI. — FIFTH SERIES. — *Preliminary.*

March 24, 1878. Jacket 5°.4 to 8°.4. Air about 6° C.

Thermometer No. 6168.	Time.	Correction.		Corrected Temperature.	Revolutions of Chronograph 1 n.	Mean Weight W.	Work per Kilogramme = 2 9.9816 W n.	Temperature.	Work per Kilogramme.	Work per Kilogramme + 2250.
		Stem.	Rad.							
75	0.9	—.003	0	1.891	3.154	8.1544	0	2	46	2296
80	1.7	....	....	2.451	6.118	8.0000	239	3	477	2727
90	3.4	—.002	—.012	3.569	12.174	8.0409	723	4	906	3156
100	5.1	....	....	4.690	18.172	8.0074	1200	5	1332	3582
110	6.8	0	—.017	5.810	24.212	7.9170	1677	6	1759	4009
120	8.5	....	....	6.936	30.397	7.8973	2161	7	2189	4439
130	10.2	+.003	—.012	8.060	36.621	7.8786	2647	8	2621	4871
140	12.0	....	....	9.190	42.854	7.8512	3132	9	3050	5300
150	13.7	+.007	+.005	10.323	49.068	7.8061	3614	10	3477	5727
160	15.5	....	....	11.459	55.898	7.7799	4108	11	3905	6155
170	17.2	+.015	+.032	12.600	61.707	7.7622	4588	12	4333	6583
180	19.0	....	....	13.742	68.036	7.7643	5078	13	4759	7009
190	20.8	+.024	+.068	14.882	74.358	7.7807	5558	14	5183	7433
200	22.6	+.028	+.092	16.025	80.716	7.8419	6047	15	5608	7858
210	24.3	....	....	17.170	87.064	7.8468	6539	16	6036	8286
220	26.1	+.039	+.150	18.316	93.402	7.8579	7030	17	6466	8716
230	27.9	....	....	19.467	99.677		7518	18	6895	9145

* The first part of the experiments were lost, as the pen of the chronograph did not work.

TABLE XLI.— Continued.

Thermometer No. 6163.	Time.	Correction.		Corrected Temperature.	Revolutions of Chronograph Σ n.	Mean Weight W.	Work per Kilogramme = Σ 9.8816 W n.	Temperature.	Work per Kilogramme.	Work per Kilogramme + 2250.
		Stem.	Rad.							
240	29.6	+.050	+.270	20.815°	105.950	7.8802	8006	19°	7320	9570
250	....	....	....	....	....	} 7.8980	....	20	7745	9995
260	....	....	....	....	....		....	21	8170	10420
270	34.9	+.069	+.351	24.072	124.863	7.9088	9482	22	8597	10847
280	36.7	....	....	25.231	131.181	7.9091	9976	23	9024	11274
290	38.5	+.087	+.450	26.395	137.560	7.8979	10474	24	9451	11701
300	40.2	....	....	27.565	143.972	7.8974	10974	25	9878	11128
310	42.1	+.109	+.583	28.748	150.467		11481	26	10305	12555
...	...	....	....	....	....	....	....	27	10733	12983
...	...	....	....	....	....	....	....	28	11160	13410

TABLE XLII.— SIXTH SERIES.

May 14, 1878. Jacket 12°.1 to 12°.4. Air about 13° C.

Thermometer No. 6165.	Time.	Correction.		Corrected Temperature.	Revolutions of Chronograph Σ n.	Mean Weight W.	Work per Kilogramme = Σ 9.9051 W n.	Temperature.	Work per Kilogramme.	Work per Kilogramme + 5432.
		Stem.	Rad.							
140	46.4	—.002	0	9.319°	1.93		0	9°	—137	5206
150	47.9	....	....	10.178	7.07	} 7.2291	870	10	293	5726
160	49.4	.000	—.007	11.032	12.19		735	11	721	6154
170	50.9	....	....	11.886	17.37	} 7.1608	1102	12	1151	6584
180	52.5	+.002	—.008	12.740	22.52		1467	13	1579	7012
190	54.0	....	....	13.596	27.70	} 7.1500	1835	14	2007	7440
200	55.5	+.006	—.002	14.454	32.88		2201	15	2434	7867
210	57.0	....	....	15.314	38.07	} 7.1512	2568	16	2863	8296
220	58.5	+.010	+.011	16.174	43.29		2938	17	8290	8723
230	60.0	....	....	17.037	48.50	} 7.1446	3306	18	3716	9149
240	61.6	+.015	+.031	17.093	53.70		3675	19	4142	9575
250	...	....	....	....	....	} 7.1536	....	20	4567	10000
260	...	....	....	....	....		....	21	4993	10426
270	66.2	+.024	+.075	20.500	69.27	} 7.1230	4778	22	5420	10853
280	67.7	....	....	21.362	74.50		5148	23	5846	11279
290	69.2	+.031	+.113	22.220	79.69	} 7.1344	5514	24	6271	11704
300	70.7	....	....	23.076	84.84		5878	25	6696	12129
310	72.2	+.039	+.158	23.928	89.97	} 7.1302	6240	26	7121	12554
320	73.7	....	....	24.774	95.05		6600	27	7547	12980
330	75.2	+.047	+.212	25.624	100.19	} 7.1117	6962	28	7973	13406
340	76.2	....	....	26.467	105.27		7319	29	8400	13833
350	78.2	+.056	+.272	27.809	110.39	} 7.0958	7680	30	8829	14262
360	79.7	....	....	28.147	115.44		8035	31	9259	14692
370	81.2	+.065	+.341	28.990	120.57	} 7.1076	8396	32	9678	15111
380	82.7	....	....	29.825	125.66		8754	33	10096	15529
390	84.2	+.076	+.417	30.663	130.78	} 7.1088	9115	..	....	....
400	85.7	....	....	31.505	135.90		9475	..	....	....
410	87.2	+.087	+.504	32.377	140.98	} 7.1064	9833	..	....	....
420	88.7	....	....	33.226	146.08		10192	..	....	....

TABLE XLIII. — Seventh Series.

May 15, 1878. Jacket 11°.8 to 12°. Air about 12° C.

Thermometer No. 6168.	Time.	Correction.		Corrected Temperature.	Revolutions of Chronograph 2 s.	Mean Weight W.	Work per Kilogramme = 29.8687 W n.	Temperature.	Work per Kilogramme.	Work per Kilogramme + 5697.
		Stem.	Rad.							
130	30.9	—.004	0	8.538	5.07		0	o	....	....
140	32.2	....	....	9.315	9.73	7.2350	335	9	199	5296
150	33.6	—.002	—.006	10.094	14.36		668	10	628	5725
160	35.0	....	....	10.875	18.98	7.3011	1003	11	1056	6153
170	36.3	0	—.010	11.654	23.56		1335	12	1484	6581
180	37.6	....	....	12.433	28.16	7.3165	1670	13	1913	7010
190	38.9	+.003	—.008	13.209	32.74		2003	14	2344	7441
200	40.2	....	....	13.984	37.31	7.3460	2337	15	2770	7867
210	41.5	+.006	—.000	14.758	41.84		2667	16	3196	8293
220	42.8	....	....	15.536	46.38	7.3094	2998	17	3623	8720
230	44.2	+.010	+.013	16.317	50.99		3332	18	4052	9149
240	45.5	....	....	17.103	55.62	7.2846	3667	19	4478	9575
250	46.9	+.014	+.032	17.891	60.29		4005	20	4906	10003
260	48.3	....	....	18.682	....	7.2822	....	21	5324	10421
270	49.6	+.019	+.056	19.475	69.63		4681	22	5754	10851
280	50.9	....	....	20.269	74.34	7.2610	5021	23	6179	11276
290	52.3	+.025	+.090	21.079	79.01		5358	24	6603	11700
300	53.6	....	....	21.866	83.71	7.2504	5697	25	7028	12125
310	55.0	+.032	+.127	22.665	88.42		6037	26	7454	12551
320	56.4	....	....	23.471	93.14	7.2893	6379	27	7883	12980
330	57.8	+.039	+.172	24.281	97.88		6722	28	8307	13404
340	59.2	....	....	25.088	102.61	7.3047	7065	29	8729	13826
350	60.5	+.046	+.222	25.896	107.36		7410	30	9157	14254
360	61.9	....	....	26.706	112.14	7.3389	7759	31	9582	14679
370	63.2	+.055	+.279	27.523	116.88		8104	32	10009	15106
380	64.6	....	....	28.346	121.62	7.4109	8454	..	....	....
390	66.0	+.065	+.345	29.172	126.34		8801	..	....	....
400	67.4	....	....	29.996	131.12	7.4356	9155	..	....	....
410	68.8	+.075	+.419	30.827	135.90	7.4581	9508	..	....	....
420	70.1	+.080	+.456	31.653	140.66		9861	..	....	....

TABLE XLIV. — Eighth Series.

May 23, 1878. Jacket 16°.2 to 16°.5. Air about 20° C.

Thermometer No. 6168.	Time.	Correction.		Corrected Temperature.	Revolutions of Chronograph 2 s.	Mean Weight W.	Work per Kilogramme = 29.9075 W n.	Temperature.	Work per Kilogramme.	Work per Kilogramme + 8409.
		Stem.	Rad.							
230	23.9	—.007	0	16.287	39.120		0	o	....	....
240	25.4	....	....	17.063	43.982	6.9137	333	17	306	8715
250	26.8	....	....	....	....	6.9358	....	18	735	9144
260	28.3	....	....	....	....		....	19	1163	9572
270	29.7	.000	+.005	19.405	58.602	6.9007	1338	20	1592	10001
280	31.2	....	....	20.190	63.503	6.9125	1673	21	2019	10428
290	32.7	....	....	20.978	68.428		2010	22	2446	10855

TABLE XLIV.—Continued.

Thermometer No. 6156.	Time	Correction. Stem	Correction. Rad.	Corrected Temperature.	Revolutions of Chronograph Σ n.	Mean Weight W.	Work per Kilogramme = Σ 9.9075 W n.	Temperature.	Work per Kilogramme.	Work per Kilogramme + 8400.
300	34.2	....	....	21.765	78.851	6.8878	2846	23°	2871	11280
310	35.6	+.008	+.040	22.554	78.283	6.8866	2682	24	3208	11707
320	37.1	....	....	23.350	88.245	6.8504	3020	25	3722	12131
330	38.6	....	....	24.151	88.314	6.8358	3363	26	4150	12559
340	40.1	+.017	+.085	24.952	98.294	6.8748	3702	27	4574	12983
350	41.6	....	....	25.751	98.275	6.9184	4044	28	4999	14408
360	43.1	....	....	26.552	103.232	6.9444	4385	29	5423	13882
370	44.6	+.028	+.144	27.361	108.216	6.9201	4727	30	5851	14260
380	46.0	....	....	28.175	113.269	6.9338	5074	31	6275	14684
390	47.5	....	....	28.989	118.281	6.9385	5418	..	....	....
400	49.0	+.039	+.217	29.800	123.329	6.9444	5766	..	....	....
410	50.6	....	....	30.624	128.899	6.9467	6115	..	....	....
420	52.1	+.047	+.281	31.445	133.480	6.9314	6464	..	....	....

TABLE XLV.—NINTH SERIES.

May 27, 1878. Jacket 19°.6 to 20°. Air about 23° C.

Thermometer No. 6163.	Time	Correction. Stem	Correction. Rad.	Corrected Temperature.	Revolutions of Chronograph Σ n.	Mean Weight W	Work per Kilogramme = Σ 9.9077 W n.	Temperature.	Work per Kilogramme.	Work per Kilogramme + 8246.
200	38.0	—.015	0	15.890	6.88		0	16°	47	8293
210	39.4	....	....	17.000	11.74	8.8108	473	17	473	8719
220	40.9	—.011	—.010	18.106	17.17		946	18	901	9147
230	42.3	....	....	19.219	22.62	8.7341	1419	19	1326	9572
240	43.8	—.005	—.011	20.829	28.18	8.6080	1895	20	1754	10000
250	45.3	....	....	21.442	33.68		2368	21	2180	10426
260	...	+.002	—.004	22.552	....	8.4800	....	22	2606	10852
270	...	....	....	23.659	....		....	23	3031	11277
280	49.8	+.009	+.012	24.771	50.55		3785	24	3457	11703
290	51.3	....	....	25.885	56.25	8.4393	4263	25	3883	12129
300	52.9	+.019	+.087	27.006	61.93		4737	26	4312	12558
310	54.4	....	....	28.183	67.63	8.4765	5215	27	4734	12980
320	56.0	+.029	+.072	29.264	73.36		5697	28	5159	13405
330	57.5	....	....	30.404	79.15	8.4552	6182	29	5584	13830
340	59.1	+.042	+.118	31.552	84.97		6669	30	6010	14256
350	60.6	....	....	32.702	90.85	8.4015	7159	31	6435	14681
360	62.2	+.056	+.178	33.853	96.78		7652	32	6860	15106
370	63.8	....	....	35.011	102.66	8.4222	8143	33	7286	15532
380	65.4	+.071	+.242	36.170	108.59		8638	34	7714	15960
390	67.0	....	....	37.831	114.45	8.4706	9128	35	8138	16384
400	68.6	+.088	+.322	38.497	120.36		9626	36	8565	16811
410	70.2	....	....	39.664	126.33	8.4316	10126	37	8988	17234
420	71.8	+.105	+.419	40.883	132.26		10620	38	9414	17660
...	...	....	....	....	....		....	39	9842	18088
...	...	....	....	....	....		....	40	10268	18514
...	...	....	....	....	....		....	41	10691	18937

TABLE XLVI. — TENTH SERIES.

June 3, 1878. Jacket 18°.1 to 18°.4. Air about 20° C.

Thermometer No. 6166.	Time.	Correction. Stem.	Correction. Rad.	Corrected Temperature.	Revolutions of Chronograph Σ n.	Mean Weight W.	Work per Kilogramme = Σ 9.8878 W n.	Temperature.	Work per Kilogramme.	Work per Kilogramme + 9076.
250	4.1	—.007	0	17.838	7.82	} 4.8899	0	18	69	9145
260	7.0	....	....	18.617	....		....	19	496	9572
270	9.9	—.003	+.004	19.401	23.19		667	20	925	10001
280	12.8	....	....	20.188	30.95	} 4.3919	1005	21	1350	10426
290	15.7	+.003	+.020	20.978	38.70		1341	22	1778	10854
300	18.7	....	....	21.763	46.41	} 4.3912	1676	23	2204	11280
310	21.6	+.008	+.037	22.551	54.21		2014	24	2627	11703
320	24.5	....	....	23.354	62.04	} 4.3907	2354	25	3054	12130
330	27.5	+.014	+.078	24.162	69.92		2696	26	3479	12555
340	30.5	....	....	24.970	77.92	} 4.3624	3041	27	3904	12980
350	33.6	+.020	+.132	25.780	85.89		3385	28	4332	13408
360	36.6	....	....	26.593	93.94	} 4.3542	3731	29	4852	13828
370	39.6	+.028	+.198	27.415	102.05		408b	30	5170	14255
380	42.7	....	....	28.246	110.34	} 4.3362	4437	31	5604	14680
390	45.8	+.036	+.281	29.079	118.49		4786	..	....	....
400	48.9	....	....	29.911	126.66	} 4.3978	5141	..	....	....
410	52.0	+.044	+.377	30.754	134.89		5499	..	....	....

TABLE XLVII. — ELEVENTH SERIES.

June 19, 1878. Jacket 19°.6 to 20°. Air about 23° C.

Thermometer No. 6163.	Time.	Correction. Stem.	Correction. Rad.	Corrected Temperature.	Revolutions of Chronograph Σ n.	Mean Weight W.	Work per Kilogramme = Σ 9.8404 W n.	Temperature.	Work per Kilogramme.	Work per Kilogramme + 10620.
250	...	—.002	0	21.450	8.933	} 6.7672	0	21	—192	10428
260	...	+.002	+.006	22.562	16.087		476	22	235	10855
270	...	....	....	....	....	} 6.7678	....	23	662	11282
280	...	+.010	+.029	24.789	30.281		1421	24	1087	11707
290	...	....	....	25.907	37.439	} 6.7749	1899	25	1511	12131
300	...	+.019	+.063	27.032	44.655		2379	26	1939	12559
310	...	....	....	28.168	51.848	} 6.7896	2860	27	2365	12985
320	...	+.031	+.113	29.307	59.098		3344	28	2789	13409
330	...	....	....	30.456	66.390	} 6.7973	3832	29	3214	13834
340	...	+.043	+.177	31.612	73.724		4323	30	3638	14258
350	...	....	....	32.774	81.153	} 6.8188	4817	31	4063	14683
360	...	+.058	+.257	33.939	88.462		5311	32	4488	15108
370	...	....	....	35.110	95.734	} 6.9165	5807	33	4913	15533
380	...	+.072	+.351	36.280	103.093		6307	34	5337	15957
390	...	....	....	37.456	110.560	} 6.7876	6808	35	5760	16380
400	...	+.087	+.463	38.637	118.121		7311	36	6187	16807
410	...	....	....	39.821	125.608	} 6.7808	7815	37	6614	17234
420	...	+.106	+.595	41.010	133.250		8321	38	7040	17660
...	...	....	....	....	....	....	....	39	7465	18085
...	...	....	....	....	....	....	....	40	7891	18511
...	...	....	....	....	....	....	....	41	8317	18937

TABLE XLVIII.—TWELFTH SERIES. EXPERIMENT OF DECEMBER 17, 1878.

Thermometer No. 6163.	Temperature of Calorimeter by No. 6163.	Temperature			Corrections				Corrected Temperature	Revolutions of Chronograph	Time	A Revolutions of Chronograph	Mean Weight W.	Work per Kilogramme		Temperature	Work per Kilogramme by Interpolation	Work per Kilogramme. + 1960
		Water Jacket	Air	Tube at bottom of 6163	Radiation	Stem Part in Air	Stem Part in Tube of Water	Total						9.85706 IV.	29.85706 W.			
70	1.248	...	8.6	4.3	0	...	0	0	1.248	5.22	14.2	6.84	7.213	486.3	0	0	–107	1853
80	2.380	3.27	...	...	–.005	...	...	–.003	2.377	12.06	16.2	6.89	7.126	484.0	486.3	1	+824	2284
90	3.503	3.30	3.7	4.2	–.003	...	0	–.005	3.408	18.95	18.8	6.89	7.104	482.5	970.3	2	755	2715
100	4.626	...	...	...	+.003	.002	...	–.002	4.624	25.84	20.3	6.89	7.104	487.1	1452.8	3	1195	3145
110	5.747	3.30	3.9	4.1	+.008	...	+.002	–.005	5.752	32.92	22.4	7.08	6.990	488.5	1939.9	4	1615	3575
120	6.868	...	...	...	+.022	+.005	...	+.014	6.882	39.92	24.4	7.00	7.080	488.9	2428.4	5	2045	4005
130	7.989	8.30	4.0	4.0	+.054	...	+.010	+.027	8.016	46.81	26.4	6.89	7.155	485.9	2914.8	6	2475	4435
140	9.110	...	...	...	+.064	+.010	...	+.044	9.154	53.70	28.4	6.89	7.193	488.5	3402.8	7	2470	4867
150	10.231	3.33	4.1	4.2	+.101	...	+.017	+.064	10.295	60.62	30.5	6.92	7.210	491.8	3894.6	8	2907	5297
160	11.352	...	...	...	+.118	+.017	...	+.090	11.442	67.51	32.5	6.89	7.168	486.8	4381.4	9	8337	5727
170	12.473	3.35	4.1	4.3	+.151	...	+.025	+.118	12.591	74.49	34.6	6.99	7.151	492.0	4878.4	10	3767	6158
180	13.592	...	...	...	+.185	+.025	...	+.151	13.743	81.48	36.6	6.99	7.187	491.7	5365.2	11	4193	6580
190	14.710	3.85	4.2	4.5	+.226	...	–.008	+.185	14.895	88.40	38.7	6.98	7.169	492.6	5857.7	12	4620	7008
200	15.825	...	...	...	+.267	+.025	...	+.226	16.051	95.44	40.7	6.98	7.185	490.9	6348.5	13	5048	7435
210	16.938	3.35	4.3	4.8	+.316	...	–.009	+.267	17.205	102.44	42.8	7.00	7.156	493.7	6842.3	14	5475	7862
220	18.047	8.40	4.5	5.2	+.366	+.034	...	+.316	18.363	109.45	44.8	7.01	7.202	497.6	7339.9	15	5902	8287
230	19.157	...	...	...	+.428	...	–.010	+.366	19.523	116.42	46.9	6.97	7.193	494.2	7834.2	16	6827	8715
240	20.265	3.50	4.6	5.6	+.425	+.046	...	+.423	20.688	123.47	48.9	7.06	7.210	501.4	8335.2	17	6755	9144
250	21.372	...	...	...	+.481	...	...	+.481	21.853	...	...	...	...	...	...	18	7184	9571
200	22.474	3.53	4.6	5.8	+.546	+.058	–.011	+.547	23.021	137.43	53.0	13.96	7.207	491.7	9326.9	19	7611	9999
270	23.572	...	...	...	+.646	...	...	+.615	24.187	144.44	55.1	7.01	7.218	498.4	9825.3	20	8039	10428
280	24.670	3.56	4.6	5.9	+.609	+.085	–.012	+.686	25.356	151.51	57.1	7.07	7.185	500.7	10323.0	21	8468	10863
...	...	...	...	...	...	...	...	...	...	...	...	...	...	...	...	22	8893	11278
...	...	...	...	...	...	...	...	...	...	...	...	...	...	...	...	23	9318	11705
...	...	...	...	...	...	...	...	...	...	...	...	...	...	...	...	24	9475	11705
...	...	...	...	...	...	...	...	...	...	...	...	...	...	...	...	25	10173	12183

* Correction for 0 point 0°.08 C. In the experiments previous to this, no correction was necessary. † These are interpolated from the observations.

TABLE XLIX. — THIRTEENTH SERIES.

Dec. 19, 1878. Jacket 3°.2 to 3°.5. Air 4°.2 to 5°.2 C.

Thermometer No. 6163.	Corrections.		Corrected Temperature.	Revolutions of Chronograph Σ n.	Mean Weight W.	Work per Kilogramme 9.8838 × Wn.	I 9 6823 Wn.	Temperature.	Work per Kilogramme.	Work +1964.
	Stem.	Rad.								
70	0	0	1.248	1.72			0	1°	—106	1858
					8.6610	485.0				
80	....	....	2.378	7.38			485.0	2	+323	2287
					8.5571	485.1				
90	0	—.003	3.500	13.11			970.1	3	754	2718
					8.4325	482.2				
100	....	....	4.626	18.89			1452.3	4	1184	3148
					8.3688	481.1				
110	+.001	+.003	5.751	24.70			1933.4	5	1612	3576
					8.4155	487.1				
120	....	....	6.881	30.55			2420.5	6	2041	4005
					8.4189	485.6				
130	+.005	+.019	8.013	36.38			2906.1	7	2472	4436
					8.3953	489.2				
140	....	....	9.148	42.27			3395.3	8	2901	4865
					8.4366	486.6				
150	+.009	+.044	10.284	48.10			3881.9	9	3331	5295
					8.4484	486.5				
160	....	....	11.424	53.92			4368.4	10	3760	5724
					8.4189	490.6				
170	+.016	+.080	12.569	59.81			4859.0	11	4187	6151
					8.3988	491.1				
180	....	....	13.713	65.72			5350.1	12	4615	6579
					8.4153	487.1				
190	+.023	+.126	14.859	71.57			5837.2	13	5045	7009
					8.3811	491.7				
200	....	....	16.005	77.50			6328.9	14	5472	7436
					8.3835	489.4				
210	+.033	+.183	17.154	83.40			6818.3	15	5898	7862
					8.3976	490.2				
220	....	....	18.300	89.30			7308.5	16	6327	8291
					8.4035	493.0				
230	+.044	+.251	19.452	95.23			7801.5	17	6758	8717
					8.4460	496.4				
240	....	....	20.604	101.17			8297.9	18	7180	9144
250	+.056	+.332	21.760	.....	} 8.4555	491.3	.....	19	7608	9572
260	....	....	22.912	112.90			9279.2	20	8038	10002
					8.4602	494.7				
270	+.069	+.424	24.065	118.81			9773.9	21	8465	10429
					8.4779	494.0				
280	....	....	25.221	124.70			10267.9	22	8891	10855
...	....	....	......	.....	......	....	......	23	9317	11281
...	....	....	......	.....	......	....	......	24	9746	11710
...	....	....	......	.....	......	....	......	25	10173	12137

TABLE L. — FOURTEENTH SERIES.

December 20, 1878. Jacket 1°.5 to 1°.9. Air about 3°.4 C.

Temperature by Kew Standard.	Time.	Corrections.			Corrected Temperature, Absolute Scale.	Revolution of Chronograph Σn	Mean Weight W.	Work per Kilogramme $= \Sigma 9.7882 W n.$	Temperature.	Work per Kilogramme.	Work per Kilogramme $+ 2210.$
		Reduction to Absolute Scale.	Stem.	Rad.							
36.0	56.0	.00	0	0	1°.82	8.03	7.3682	0	2°	77	2287
38.5	58.4	....	....	....	3.23	16.37	7.3458	601	3	503	2713
41.0	.9	—.01	.00	+.01	4.62	24.78	7.3705	1206	4	936	3146
43.5	3.3	....	....	....	6.02	33.19	7.4012	1812	5	1370	3580
46.0	5.8	—.02	+.01	+.04	7.43	41.48	7.4142	2412	6	1803	4013
48.5	8.2	....	....	....	8.84	49.81	7.4177	3016	7	2226	4436
51.0	10.7	—.03	+.02	+.09	10.26	58.18	7.4300	3624	8	2656	4866
53.5	13.2	....	....	....	11.68	66.56	7.4107	4234	9	3084	5294
56.0	15.6	—.04	+.03	+.16	13.12	74.95	7.3493	4842	10	3513	5723
58.5	18.2	....	....	....	14.56	83.56	7.3269	5461	11	3942	6152
61.0	20.7	—.04	+.05	+.25	16.01	92.27	7.2335	6085	12	4369	6579
63.5	23.3	....	....	....	17.46	100.09	7.1603	6703	13	4790	7000
66.0	25.9	—.05	+.06	+.38	18.92	109.95	7.2075	7330	14	5220	7430
68.5	28.5	....	....	....	20.39	118.84	7.1839	7957	15	5650	7860
71.0	31.2	—.05	+.08	+.52	21.86	127.83	7.2122	8589	16	6081	8291
73.5	33.8	....	....	.....	23.34	136.75	7.2252	9218	17	6507	8717
76.0	36.5	—.05	+.10	+.69	24.84	145.78	7.2134	9857	18	6935	9145
78.5	39.2	....	....	....	26.33	154.80	.....	10493	19	7364	9574
....	....	....	....	....	....	.....	.....	.....	20	7791	10001
....	....	....	....	....	....	.....	.....	.....	21	8219	10429
....	....	....	....	....	....	.....	.....	.....	22	8648	10858
....	....	....	....	....	....	.....	.....	.....	23	9074	11284
....	....	....	....	....	....	.....	.....	.....	24	9499	11709
....	....	....	....	....	....	.....	.....	.....	25	9925	12135
....	....	....	....	....	....	.....	.....	.....	26	10352	12562

190   PROCEEDINGS OF THE AMERICAN ACADEMY

TABLE LI. — Work in Kilogramme-Meters at Baltimore to heat One Kilogramme of Water from an Unknown Point to a Given Temperature on the Absolute Scale.

Temperature	Preliminary Results with simple Tin Jacket for the Radiation.				March 24.		Final results with the Water Jacket around the Calorimeter to make the Radiation definite.								Mean of Preliminary Results.	Mean of Final Results.	General Mean of all Results.	General Mean from Combination of Table by Parts.
	Jan. 16. 6168. Weight 7.4 kil. 36° per hour.	Mar. 7. 6168. Weight 7.6 kil. 39° per hour.	March 12. 6166. Weight 7.6 kil. 38° per hour.	1 Series. 6168. Weight 7.4 kil. 36° per hour.	2 Series. 6168. Weight 7.8 kil. 38° per hour.	May 14 6165. Weight 7.1 kil. 34° per hour.	May 16. 6166. Weight 6.8 kil. 36° per hour.	May 23. 6166. Weight 6.9 kil. 32° per hour.	May 27. 6168. Weight 8.6 kil. 42° per hour.	June 8. 6166. Weight 4.4 kil. 16° per hour.	June 19 6163. Weight 6.8 kil. 31° per hour.	Dec. 17. 6168. Weight 7.3 kil. 34° per hour.	Dec. 19. 6163. Weight 8.4 kil. 43° per hour.	Dec. 20. 6168. Weight 7.8 kil. 34° per hour.				
2	…	…	…	…	2206	…	…	…	…	…	…	2284	2287	2287	…	2286	2288	2289
3	…	…	…	…	2727	…	…	…	…	…	…	2715	2718	2718	…	2715	2718	2719
4	…	…	…	…	3156	…	…	…	…	…	…	3145	3148	3146	…	3146	3149	3150
5	…	…	…	…	3582	…	…	…	…	…	…	3575	3576	3580	…	3577	3578	3579
6	…	…	…	…	4009	…	…	…	…	…	…	4006	4005	4013	…	4008	4008	4009
7	…	…	…	…	4489	…	…	…	…	…	…	4439	4436	4436	…	4437	4437	4438
8	…	…	…	…	4871	…	…	…	…	…	…	4867	4865	4866	4871	4866	4868	4868
9	…	…	…	5300	5300	5296	5296	…	…	…	…	5297	5295	5294	5300	5296	5297	5297
10	5728	…	…	5726	5727	5726	5725	…	…	…	…	5727	5724	5728	5727	5725	5726	5725
11	6156	…	…	6154	6155	6154	6153	…	…	…	…	6153	6151	6152	6155	6153	6153	6152
12	6582	…	…	6582	6588	6584	6581	…	…	…	…	6580	6579	6579	6582	6581	6581	6580
13	7009	7010	…	7009	7009	7012	7010	…	…	…	…	7008	7009	7000	7009	7008	7008	7007
14	7436	7437	…	7436	7433	7440	7441	…	…	…	…	7435	7436	7430	7436	7436	7436	7435
15	7864	7864	7868	7862	7858	7867	7867	…	…	…	…	7862	7862	7860	7863	7864	7863	7862
16	8292	8292	8295	8289	8296	8296	8293	…	8293	…	…	8287	8291	8291	8291	8292	8291	8290

17	8720	8721	8721	8717	8716	8728	8720	8716	8719			8716	8717	8717	8719	8718	8718	8718
18	9147	9145	9147	9147	9145	9149	9149	9144	9147	9145		9144	9144	9146	9146	0146	9146	9146
19	9573	9573	9674	9574	9570	9576	9575	9672	9572	9572		9571	9572	9574	9573	9578	9573	9573
20	9099	10001	10000	10000	9995	10000	10008	10001	10000	10001		9999	10002	10001	9099	10001	10000	10000
21	10428	10427	10427	10426	10420	10426	10421	10428	10426	10426	10428	10428	10429	10429	10426	10427	10426	10426
22	10852	10853	10852	10852	10847	10853	10851	10855	10852	10854	10855	10853	10855	10868	10861	10854	10858	10853
23	11279	11279	11275	11278	11274	11279	11276	11290	11277	11280	11282	11278	11281	11284	11277	11280	11279	11279
24	11706	11704	11700	11704	11701	11704	11700	11707	11708	11703	11707	11706	11710	11709	11708	11705	11706	11706
25	12133	12130	12125	12180	12128	12129	12125	12131	12129	12130	12181	12188	12187	12186	12129	12131	12130	12130
26	12560	12556	12560	12563	12555	12564	12551	12559	12568	12555	12559			12562	12555	12557	12556	12557
27		12980	12977		12983	12980	12980	12983	12980	12980	12985				12980	12981	12981	12982
28		13405	13402		13410	13406	18404	13408	13405	13408	13409				13406	13407	13406	13407
29		13829	18825			13833	13826	13832	13830	13828	13834				13827	13829	13828	13829
30		14253	14252			14262	14264	14260	14256	14255	14258				14258	14257	14256	14257
31		14679	14677			14692	14679	14684	14681	14680	14683				14678	14683	14681	14682
32		15106				15111	15106		15106		15108					15108	15106	15104
33		15534				15529			15532		15583					15531	15532	15530
34		15961							15960		15957					16059	15959	15956
35		16389							16384		16380					16382	16384	16381
36		16816							16811		16807					16809	16811	16808
37		17242							17234		17234					17234	17287	17284
38									17660		17660					17660	17660	17660
39									18088		18085					18086	18086	18086
40									18514		18511					18512	18512	18512
41									18937		18937					18937	18937	18937

TABLE LII.—Mechanical Equivalent of Heat in Kilogramme-Meters at Baltimore, each value calculated from a Rise of 10° C. in Temperature.

Temp. C.	Jan. 16	March 7	March 12	March 24 1st Ser.	March 24 2d Ser.	May 14	May 15	May 23	May 27	June 3	June 19	Dec. 17	Dec. 19	Dec. 20	General Mean	From regular Curve
4°	…	…	…	…	…	…	…	…	…	…	…	…	…	…	…	…
5	…	…	…	…	…	…	…	…	…	…	…	…	…	…	…	429.7
6	…	…	…	…	…	…	…	…	…	…	…	…	…	…	…	429.4
7	…	…	…	…	428.7	…	…	…	…	…	…	429.6	429.2	429.2	429.1	429.2
8	…	…	…	…	428.2	…	…	…	…	…	…	429.3	429.1	428.7	428.8	428.9
9	…	…	…	…	427.7	…	…	…	…	…	…	429.0	428.8	428.4	428.5	428.6
10	…	…	…	…	427.6	…	…	…	…	…	…	428.7	428.6	428.0	428.2	428.4
11	…	…	…	…	427.7	…	…	…	…	…	…	428.1	428.6	427.8	428.0	428.2
12	…	…	…	…	427.7	…	…	…	…	…	…	427.6	428.1	428.1	427.9	428.0
13	…	…	…	427.5	427.4	…	…	…	…	…	…	427.7	427.9	427.9	427.7	427.7
14	…	…	427.9	427.4	427.0	427.9	427.9	…	…	…	…	427.4	427.7	428.0	427.6	427.5
15	…	…	…	427.5	426.8	427.4	427.8	…	…	…	…	427.2	427.8	427.8	427.4	427.3
16	427.1	…	…	427.2	426.5	427.2	426.8	…	…	…	…	427.5	427.8	427.7	427.2	427.1
17	427.3	…	…	427.0	426.4	426.9	427.0	…	…	…	…	427.3	427.6	427.9	427.1	426.9
	427.0															

	18	19	20	21	22	23	24	25	26	27	28	29	30	31	32	33	34	35	36
	426.7	426.5	426.4	426.2	426.1	425.9	425.8	425.7	425.6	425.5	425.4	425.4	425.4	425.4	425.4	425.4	425.5	425.5	425.5
	427.0	426.9	426.7	426.4	426.1	426.0	425.6	425.5	425.6	425.4	425.5	425.4	425.4	425.4	425.3	425.5	425.3	425.6	425.5
	428.4	427.0	427.5	427.1	…	…	…	…	…	…	…	…	…	…	…	…	…	…	…
	427.1	427.4	427.5	…	…	…	…	…	…	…	…	…	…	…	…	…	…	…	…
	427.0	427.0	427.1	…	…	…	…	…	…	…	…	…	…	…	…	…	…	…	…
	…	…	…	…	…	…	…	…	425.5	425.8	425.1	425.0	424.9	424.8	424.9	425.1	425.1	425.3	425.4
	…	…	…	…	…	…	426.3	425.6	425.4	425.4	…	…	…	…	…	…	…	…	…
	…	…	426.5	426.1	425.8	425.8	425.6	425.5	425.4	425.5	425.7	425.6	425.8	425.4	425.5	425.8	425.8	425.6	…
	…	…	…	…	426.8	426.4	426.0	425.9	425.6	…	…	…	…	…	…	…	…	…	…
	426.6	425.9	425.8	425.8	426.0	425.5	425.1	425.1	425.8	425.5	…	…	…	…	…	…	…	…	…
	426.7	426.4	426.2	425.8	425.7	425.7	425.8	425.2	425.6	425.8	425.0	…	…	…	…	…	…	…	…
	426.5	426.8	427.0	426.9	426.7	426.5	…	…	…	…	…	…	…	…	…	…	…	…	…
	426.9	426.8	426.8	426.4	…	…	…	…	…	…	…	…	…	…	…	…	…	…	…
	…	…	425.7	425.5	425.6	425.5	425.1	425.2	425.0	…	…	…	…	…	…	…	…	…	…
	425.9	426.7 / 426.0	426.0	426.4	425.9	426.0	425.6	425.2	425.2	425.3	425.5	425.7	425.9	426.0	426.2	…	…	…	…
	427.0	427.0	426.9	426.8	…	…	…	…	…	…	…	…	…	…	…	…	…	…	…

VOL. XV. (N. S. VII.)   13

194   PROCEEDINGS OF THE AMERICAN ACADEMY

TABLE LIII.

Mechanical Equivalent of Heat. 10° Series on the

Temperature						Work.				Absolute Thermometric Scale.				Mercurial Thermometric Scale, the Glass similar to the			
Absolute Scale.	Absolute Temp., using $m = .0018$.	Approximate, Mercurial Thermometer.				Per Kilogr. of Water				Kilogr.-Meters at Baltimore.	Using $m = .0018$.	Using $m = .0018$.	Using $m = .0005$.	Geissler Standard.	New Standard.	Baudin 7316 or 7834.	Fastré.
* Absolute Temperature, using $m = .0005$.		Baudin 6167.	Geissler Standard.	New Standard.	Baudin 7316 or 7834.	Fastré.	Kilogrammes-Meters at Baltimore, B.	Kilogr.-Meters at Paris, B = 1.00086.	Kilogr.-Meters at Berlin, B = 1.00117.	Per Gramme of Water on the C. G. S. System.		Absolute C.G.S. System.					
0.2	2.002	2°.03	2°.03	2°.01	2°.02	2°.01	2289	2295	2298	.00000.	....	....	....	....	....	....	....
3	3.003	3.04	3.04	3.01	3.02	3.01	2719	2725	2727	2443	....	....	....	....	....	....	....
4	4.004	4.05	4.05	4.01	4.03	4.02	3150	3156	3158	2864	....	....	....	....	....	....	....
5	5.005	5.06	5.07	5.02	5.04	5.02	3579	3584	3587	3287	....	....	....	....	....	....	....
6	6.006	6.08	6.09	6.02	6.05	6.03	4009	4014	4016	3707	429.7	429.4	4211	4208	423.6	428.3	426.2	427.7
7	7.006	7.09	7.10	7.02	7.06	7.03	4438	4443	4444	4129	429.4	429.2	4208	4206	423.5	428.0	426.1	427.5
8	8.007	8.10	8.11	8.02	8.06	8.03	4868	4872	4874	4549	429.2	429.0	4206	4204	423.5	427.8	426.0	427.3
9	9.007	9.11	9.12	9.02	9.07	9.04	5297	5301	5302	4970	428.9	428.7	4203	4201	423.3	427.6	425.8	427.1
10	10.008	10.12	10.13	10.03	10.07	10.04	5726	5729	5730	5391	428.6	428.4	4201	4199	423.2	427.4	425.6	426.8
11	11.008	11.13	11.14	11.03	11.08	11.04	6152	6155	6156	5810	428.4	428.2	4199	4197	423.2	427.2	425.5	426.7
12	12.008	12.14	12.15	12.03	12.08	12.05	6580	6583	6584	6229	428.2	428.0	4197	4195	423.1	427.0	425.4	426.5
13	13.009	13.15	13.16	13.04	13.09	13.05	7007	7010	7010	6648	428.0	427.9	4195	4194	423.1	426.8	425.3	426.4
14	14.009	14.16	14.18	14.04	14.10	14.05	7436	7437	7438	7067	427.7	427.6	4192	4191	423.0	426.6	425.1	426.1
15	15.009	15.17	15.18	15.04	15.10	15.05	7862	7864	7864	7486	427.5	427.4	4190	4189	423.0	426.5	425.0	426.0
16	16.010	16.19	16.19	16.04	16.10	16.06	8290	8291	8292	7905	427.3	427.2	4188	4187	423.0	426.3	424.9	425.9
17	17.010	17.20	17.21	17.04	17.11	17.06	8718	8719	8719	8324	427.1	427.0	4186	4185	423.0	426.2	424.8	425.7
										8744	426.9	426.8	4184	4183	423.0	426.0	424.7	425.6

18	18.010	18.21	18.22	18.05	18.12	18.07	9146	9147	9147	9163	426.7	426.7	4182	4182	422.9	425.8	424.6	425.5
19	19.010	19.21	19.23	19.05	19.13	19.07	9673	9674	9574	9582	426.5	426.5	4180	4180	422.9	425.7	424.5	425.3
20	20.010	20.22	20.23	20.05	20.13	20.07	10000	10000	10000	10000	426.4	426.4	4179	4179	423.0	425.6	424.5	425.3
21	21.010	21.22	21.24	21.06	21.13	21.07	10426	10426	10425	10118	426.2	426.2	4177	4177	423.0	425.5	424.4	425.1
22	22.010	22.22	22.24	22.06	22.13	22.07	10853	10852	10852	10836	426.1	426.1	4176	4176	423 0	425.4	424.4	425.1
23	23.010	23.23	23.25	23.06	23.14	23.08	11279	11278	11278	11253	425.9	426.0	4174	4176	423.0	425.2	424.3	424.9
24	24.010	24.23	24.25	24.06	24.14	24.06	11706	11705	11704	11672	425.8	425.9	4178	4174	423.0	425.2	424.2	424.9
25	25.009	25.24	25.26	25.06	25.14	25.08	12130	12128	12128	12088	425.7	425.8	4172	4178	423.1	425.1	424.2	424.8
26	26.009	26.25	26.26	26.06	26.15	26.08	12537	12555	12564	12506	425.6	425.7	4171	4172	423.2	425.0	424.2	424.8
27	27.008	27.25	27.27	27.06	27.16	27.08	12982	12979	12979	12922	425.5	425.6	4170	4171	423.2	425.0	424.2	424.7
28	28.008	28.25	28.27	28.06	28.15	28.08	13407	13404	13403	13339	425.5	425.6	4170	4171	423.4	425.0	424.8	424.8
29	29.008	29.26	29.29	29.06	29.16	29.09	13829	13826	13826	13763	425.4	425 6	4169	4171	423.4	424.9	424.3	424.7
30	30.007	30.26	30.20	30.06	30.16	30.09	14257	14253	14252	14172	425.4	425.6	4169	4171	423.6	425.0	424.4	424.8
31	31.007	31.27	31.30	31.06	31.16	31.09	14682	14678	14677	14588	425.4	425 6	4169	4171	423.7	425.0	424.5	424.8
32	32.006	32.27	32.80	32.06	32.16	32.09	15104	15100	15098	15002	425.4	425.6	4169	4171	423.9	425.0	424.5	424.9
33	33.005	33.28	33.31	33.07	33.17	33.09	15530	15525	15524	15420	425.4	425.7	4170	4172	424.0	426.1	424.6	424.9
34	34.005	34.28	34.31	34.07	34.17	34.09	15956	15951	15949	15837	425.5	425.7	4170	4172	424.2	425.2	424.8	425.1
35	35.004	35.28	35.31	35.07	35.17	35.09	16381	16376	16373	16264	425.5	425.8	4170	4173	424.4	425.2	424.9	425.1
36	36.003	36.29	36.82	36.07	36.17	36.10	16808	16802	16800	16672	425.5	425 8	4170	4173	424.6	425.8	425.0	425 2
37	37.003	37.29	37.32	37.07	37.17	37.10	17234	17228	17226	17090								
38	38.002	38.29	38.32	38.07	38.17	38.10	17661	17654	17652	17508								
39	39.001	39.29	39.82	39.07	39.17	39.10	18086	18079	18077	17925								
40	40.000	40.29	40.32	40.07	40.17	40.10	18512	18505	18502	18342								
41	40.999	41.30	41.83	41.07	41.18	41.10	18937	18929	18927	18759								

* See Appendix to Thermometry.

TABLE LIV. — FINAL MOST PROBABLE RESULTS.

Temperature on the Absolute Scale. $m = .00015$.	Work.		Mechanical Equivalent.		Temperature on the Absolute Scale. $m = .00015$.	Work.		Mechanical Equivalent.	
	Kilogramme-Meters at Baltimore.	Absolute C. G. S. System.	Kilogramme-Meters at Baltimore.	1° Series. Absolute C. G. S. System.		Kilogramme-Meters at Baltimore.	Absolute C. G. S. System.	Kilogramme-Meters at Baltimore.	1° Series. Absolute C. G. S. System.
		00000.		0000.			00000.		0000.
2	2289	2443	....	....	22	10852	10835	426.1	4176
3	2720	2865	....	....	23	11278	11253	426.0	4175
4	3150	3286	....	....	24	11704	11670	425.9	4174
5	3580	3708	429.8	4212	25	12130	12088	425.8	4173
6	4009	4129	429.5	4209	26	12556	12505	425.7	4172
7	4439	4550	429.3	4207	27	12982	12922	425.6	4171
8	4868	4970	429.0	4204	28	13407	13339	425.6	4171
9	5297	5390	428.8	4202	29	13833	13756	425.5	4170
10	5726	5811	428.5	4200	30	14258	14173	425.6	4171
11	6154	6230	428.3	4198	31	14684	14590	425.6	4171
12	6582	6650	428.1	4196	32	15110	15008	425.6	4171
13	7010	7070	427.9	4194	33	15535	15425	425.7	4172
14	7438	7489	427.7	4192	34	15961	15842	425.7	4172
15	7865	7908	427.4	4189	35	16387	16259	425.8	4173
16	8293	8327	427.2	4187	36	16812	16676	425.8	4173
17	8720	8745	427.0	4185	37	17238	17094	....	....
18	9147	9164	426.8	4183	38	17664	17511	....	....
19	9574	9582	426.6	4181	39	18091	17930	....	....
20	10000	10000	426.4	4179	40	18517	18347	....	....
21	10426	10418	426.2	4177	41	18943	18765	....	....

TABLE LV. — QUANTITY TO ADD TO THE EQUIVALENT AT BALTIMORE TO REDUCE TO ANY LATITUDE.

Latitude.	Addition in Kilogramme-Meters.
0°	+ 0.89
10	+ 0.82
20	+ 0.63
30	+ 0.34
40	+ 0.08
50	− 0.41
60	− 0.77
70	− 1.06
80	− 1.26
90	− 1.33

Manchester —0.5; Paris —0.4; Berlin —0.5.

V. CONCLUDING REMARKS, AND CRITICISM OF RESULTS AND METHODS.

On looking over the last four columns of Table LIII., which gives the results of the experiments as expressed in terms of the different mercurial thermometers, we cannot but be impressed with the unsatisfactory state of the science of thermometry at the present day, when nearly all physicists accept the mercurial thermometer as the standard between 0° and 100°. The wide discrepancy in the results of calorimetric experiments requires no further explanation, especially when physicists have taken no precaution with respect to the change of zero after the heating of the thermometer. They show that thermometry is an immensely difficult subject, and that the results of all physicists who have not made a special study of their thermometers, and a comparison with the air thermometer, must be greatly in error, and should be rejected in many cases. And this is specially the case where Geissler thermometers have been used.

The comparison of my own thermometers with the air thermometer is undoubtedly by far the best so far made, and I have no improvements to offer beyond those I have already mentioned in the "Appendix to Thermometry." And I now believe that, with the improvement to the air thermometer of an artificial atmosphere of constant pressure, we could be reasonably certain of obtaining the temperature at any point up to 50° C. within 0.°01 C. from the mean of two or three observations. I believe that my own thermometers scarcely differ much more than that from the absolute scale at any point up to 40° C., but they represent the mean of eight observations. However, there is an uncertainty of 0.°01 C. at the 20° point, owing to the uncertainty of the value of m. But taking $m = .00015$, I hardly think that the point is uncertain to more than that amount for the thermometers Nos. 6163, 6165, and 6166.

As to the comparison of the other thermometers, it is evidently unsatisfactory, as they do not read accurately enough. However, the figures given in Table LIII. are probably very nearly correct.

The study of the thermometers from the different makers introduces the question whether there are *any* thermometers which stand below the air thermometer between 0 and 100°. As far as I can find, nobody has ever published a table showing such a result, although Boscha *infers* that thermometers of "Cristal de Choisy-le-Roi" *should* stand below, and his inference has been accepted by Regnault. But it does not seem to have been proved by direct experiment. My

Baudin thermometers seem to contain lead as far as one can tell from the blackening in a gas flame, but they stand very much above the air thermometer at 40°. I have since tried some of the Baudin thermometers up to 300°, and find that they stand *below* the air thermometer between 100° and 240°; they coincide at *about* 240°, and stand above between 240° and 300°. This is very nearly what Regnault found for "Verre Ordinaire." It is to be noted that the formula obtained from experiments below 100° makes them coincide at 233°, which is remarkably close to the result of actual experiment, especially as it would require a long series of experiments to determine the point within 10°.

The comparison of thermometers also shows that all thermometers in accurate investigations should be used as thermometers with arbitrary scales, neither the position of the zero point nor the interval between the 0° and 100° points being assumed correct. The text books only give the correction for the zero point, but my observations show that the interval between the 0° and 100° points is also subject to a secular change as well as to the temporary change due to heating. Of all the thermometers used, the Geissler is the worst in this as in other respects, except accuracy of calibration, in which it is equal to most of the others.

The experiments on the specific heat of water show an undoubted *decrease* as the temperature rises, a fact which will undoubtedly surprise most physicists as much as it surprised me. Indeed, the discovery of this fact put back the completion of this paper many months, as I wished to make certain of it. There is now no doubt in my mind, and I put the fact forth as proved. The only way in which an error accounting for this decrease could have been made appears to me to be in the determination of m in "Thermometry." The determination of m rests upon the determination of a difference of only 0.°05 C. between the air thermometer and the mercurial, the 0° and 40° points coinciding, and also upon the comparison of the thermometers with others whose value of m was known, as in the Appendix. Although the quantity to be measured is small, yet there can be no doubt at least that m is larger than zero; and if so, the specific heat of water certainly has a minimum at about 30°.

One point that might be made against the fact is that the Kew standard, Table L., gives less change than the others. But the calibration of the Kew standard, although excellent, could hardly be trusted to 0°.02 or 0°.03 C., as the graduation was only to $\frac{1}{2}$° F. In drawing the curve for the difference between the Kew standard and the air thermometers, I ignored small irregularities and drew a regular curve. On looking over the observations again, I see that, had I taken account of the small irregularities, it would have made the observations agree more nearly with the other thermometers. Hence the objection vanishes. However, I intend working up some observations which I have with the Kew standard at a higher temperature, and shall publish them at a future time.

There is one other error that might produce an apparent decrease in the specific heat, and that is the slight decrease in the torsion weight from the beginning to the end of most of the experiments, probably due to the slowing of the engine. By this means the torsion circle might lag behind. I made quite an investigation to see if this source of error existed, and came to the conclusion that it produced no perceptible effect. An examination of the different experiments shows this also; for in some of them the weight increases instead of decreases. See Tables XXXVII. to L.

The error from the formation of dew might also cause an apparent decrease; but I have convinced myself by experiment, and others can convince themselves from the tables, that this error is also inappreciable.

The observations seem to settle the point with regard to the specific heat at the 4° point within reasonable limits. There does not seem to be a change to any great extent at that point, but the specific heat decreases continuously through that point. It would hardly be possible to arrive at this so accurately as I have done by any method of mixture, for Pfaundler and Platter, who examined this point, could not obtain results within one per cent, while mine show the fact within a fraction of one per cent.

The point of minimum cannot be said to be known, though I have placed it provisionally between 30° and 35° C., but it may vary much from that.

The method of obtaining the specific heat of the calorimeter seems to be good. The use of solder introduces an uncertainty, but it is too small to affect the result appreciably. The different determinations of the specific heat of the calorimeter do not agree so well as they might; but the error in the equivalent resulting from this error is very small, and, besides, the mean result agrees well with the calculated result. It may be regarded as satisfactory.

The apparatus for determining the equivalent could scarcely be improved much, although perhaps the record of the torsion might be made automatic and continuous. The experiment, however, might be improved in two ways; first, by the use of a motive power more regular in its action ; and, second, by a more exact determination of the loss due to radiation. The effect of the irregularity of the engine has been calculated as about 1 in 4,000, and I suppose that the error due to it cannot be as much as that after applying the correction. The error due to radiation is nearly neutralized, at least between 0° and 30°, by using the jacket at different temperatures. There may be an error of a small amount at that point (30°) in the direction of making the mechanical equivalent too great, and the specific heat may keep on decreasing to even 40°.

Between the limits of 15° and 25° I feel almost certain that no subsequent experiments will change my values of the equivalent so much as two parts in one thousand, and even outside those limits, say between 10° and 30°, I doubt whether the figures will ever be changed much more than that amount.

It is my intention to continue the experiments, as well as work up the remainder of the old ones. I shall also use some liquids in the calorimeter other than water, and so have the equivalent in terms of more than one fluid.

BALTIMORE, 1878-79.   Finished May 27, 1879.

NOTE FROM PROFESSOR ROWLAND.

[*Comparison with Dr. Joule's Thermometer.*]

Dr. Joule has kindly sent me the comparison of my thermometer, No. 6166, with his, and the result will be published in full in the "Proceedings of the American Academy of Sciences." In this manner I have been able to make the exact reduction of his results to the air thermometer. The following are the results:

DATE	METHOD	Temperature of water	Joule's value of the equivalent	Joule's value reduced to the air thermometer, and to the latitude of Baltimore.		Rowland's value	Difference
				English System	Metric System		
1847	Friction of water...	15°.	781.5	787.0	442.8	427.4	+15.4
1850	" " " ..	14°.	772.7	778.0	426.8	427.7	− .9
"	" " mercury	9°.	772.8	779.2	427.5	428.8	− 1.3
"	" " "	9°.	775.4	781.4	428.7	428.8	− .1
"	" " iron	9°.	776.0	782.2	429.1	428.8	+ .3
"	" " "	9°.	773.9	780.2	428.0	428.8	− .8
1867	Electric heating...	18°.6			428.0	426.7	+ 1.3
1878	Friction of water...	14°.7	772.7	776.1	425.8	427.6	− 1.8
"	" " " ..	12°.7	774.6	778.5	427.1	428.0	− .9
"	" " " ..	15°.5	778.1	776.4	426.0	427.3	− 1.3
"	" " " .	14°.5	767.0	770.5	422.7	427.5	− 4.8
"	" " " ..	17°.3	774.0	777.0	426.3	426.9	− .6

The mean difference of the two amounts to only 1 in 430, almost exactly the same as I estimated in the body of my paper, an extremely satisfactory result.

H. A. ROWLAND.

BALTIMORE, February 16th, 1880.

www.ingramcontent.com/pod-product-compliance
Lightning Source LLC
Chambersburg PA
CBHW030901170426
43193CB00009BA/699